Whose Story Wins

Rise of the Noosphere, Noopolitik, and Information-Age Statecraft

DAVID RONFELDT, JOHN ARQUILLA

July 2020

For more information on this publication, visit www.rand.org/t/PEA237-1

Library of Congress Cataloging-in-Publication Data is available for this publication.
ISBN: 978-1-9774-0530-2

Published by the RAND Corporation, Santa Monica, Calif.
© Copyright 2020 RAND Corporation
RAND® is a registered trademark.

Cover image: Siarhei / Adobe Stock.

Support RAND
Make a tax-deductible charitable contribution at
www.rand.org/giving/contribute

www.rand.org

Preface

In this Perspective, the authors urge strategists to consider a new concept for adapting U.S. grand strategy to the information age—*noopolitik*, which favors the use of "soft power"—as a successor to realpolitik, with its emphasis on "hard power." The authors illuminate how U.S. adversaries are already deploying dark forms of noopolitik (e.g., weaponized narratives, strategic deception, epistemic attacks); propose ways to fight back; and discuss how the future of noopolitik might depend on what happens to the *global commons*—i.e., the parts of Earth that fall outside national jurisdictions and to which all nations are supposed to have access.

The authors expand on many of the ideas they first proposed in a 1999 RAND Corporation report titled *The Emergence of Noopolitik: Toward an American Information Strategy*, in which they described the emergence of a new globe-circling realm: the *noosphere*. First, Earth developed a *geosphere*, or a geological mantle; second, a *biosphere*, consisting of plant and animal life. Third to develop will be Earth's *noosphere*, a global "thinking circuit" and "realm of the mind" upheld by the digital information revolution. As the noosphere expands, it will profoundly affect statecraft—the conditions for traditional realpolitik strategies will erode, and the prospects for noopolitik strategies will grow. Thus, the decisive factor in today's and tomorrow's wars of ideas is bound to be "whose story wins"—the essence of noopolitik. To improve prospects for the noosphere and noopolitik, U.S. policy and strategy should, among other initiatives, treat the global commons as a pivotal issue area, uphold "guarded openness" as a guiding principle, and institute a new requirement for periodic reviews of the U.S. "information posture."

The RAND Corporation is a research organization that develops solutions to public policy challenges to help make communities throughout the world safer and more secure, healthier and more prosperous. RAND is nonprofit, nonpartisan, and committed to the public interest.

This research was funded by income earned on client-funded research and through the support of RAND donors. Philanthropic contributions support our ability to take the long view, tackle tough and often-controversial topics, and share our findings in innovative and compelling ways. RAND's research findings and recommendations are based on data and evidence and therefore do not necessarily reflect the policy preferences or interests of its clients, donors, or supporters.

This research was conducted within the International Security and Defense Policy Center of the RAND National Security Research Division (NSRD). NSRD conducts research and analysis for the Office of the Secretary of Defense, the Joint Staff, the Unified Combatant Commands, the defense agencies, the Navy, the Marine Corps, the U.S. Coast Guard, the U.S. Intelligence Community, allied foreign governments, and foundations.

For more information on the RAND International Security and Defense Policy Center, see www.rand.org/nsrd/isdp or contact the director (contact information is provided on the webpage).

Contents

Table

Summary

U.S. grand strategy is not adapting well to the information age; a rethinking is needed. For decades, countless writings have pointed this out—ours among them—and some marginal improvements have occurred. But it is time to urge a deeper rethinking in light of new threats facing American society and its institutions. The United States is neither countering these threats effectively nor looking ahead in the best ways possible.

The problem is not so much technological in nature—the United States has excellent information, communications, and sensing technologies. Rather, it is mainly a cognitive challenge. Adversaries—both nations and nonstate networks—are using dark new modes of political, social, cultural, and psychological warfare against the United States, its allies, and its friends: wars of ideas, battles of stories, weaponized narratives, memetic operations, and epistemic attacks, all deliberately designed to undo strengths and exploit weaknesses.

Strategists of all stripes—theorists and practitioners—remain unsettled and often baffled about how best to analyze, organize, and act amid this stormy flux. All of the auguries suggest that matters might grow worse before they turn better in the years ahead.

The most advisable way ahead for the United States, given its prominent role in world affairs, is to reposition statecraft and grand strategy by merging two streams of thought: the first involves the well-known distinction between hard power and soft power, the second a lesser-known distinction about the geosphere, biosphere, and noosphere. At first glance, the two streams might seem unrelated, but they are starting to come together in ways that should be recognized—and the sooner, the better. Doing so reveals a new kind of information-age statecraft that we call *noopolitik* as a successor to traditional realpolitik.

Hard Power Versus Soft Power

Strategists have traditionally thought and planned primarily in terms of tangible, material, "hard" forms of power—military forces, economic capabilities, natural resources—and they developed realpolitik to express their hard-power dispositions as a mode of statecraft that emphasizes seeking relative advantages through displays of force. (See Table 5.1 for more information about realpolitik.) A realization that immaterial, ideational, "soft" forms of power—ideas, values, norms, battles for hearts and minds—matter as profoundly as hard forms of power started to take hold in the early 1990s, when the end of the Cold War and peaceful dissolution of the Soviet Union bore out the potential of ideational approaches to statecraft. Hard power played crucial roles in deterrence and containment from the 1940s to the 1980s, but it was soft power that brought the decades of high-stakes confrontations to a conclusion. Moreover, by then, the internet and other digital information technologies were on the rise, and strategists were beginning to view information itself as a new form of power, one that favored the soft side of the spectrum.

However, the American idea of soft power contained unnoticed flaws. The original definition tended to treat soft power as good and hard power as bad or at least mean-spirited—soft power was said to be fundamentally about persuasive attraction, hard power about coercion. But in actuality, soft power is not just about beckoning in attractive, upbeat, moralistic ways that make the United States and its allies look good. It can be wielded in tough, dark, heavy ways, too, as in efforts to warn, embarrass, denounce, disinform, deceive, shun, or repel a target. Moreover, soft power does not inherently favor the "good guys"; malevolent leaders—e.g., an Adolf Hitler, an Osama bin Laden, or various authoritarian leaders today—often prove eager and adept at using soft-power measures in their efforts to dominate at home and abroad.

And this is what has been happening, in numerous countries around the world. Adversaries—from nation-state actors in China, Iran, and Russia to such nonstate networks as al Qaeda, the Islamic State (IS), and WikiLeaks—have been quickly learning to develop dark approaches to soft power, especially online, in order to undermine American society and other liberal societies and challenge their positions in the world. Moscow has fielded new narratives to extol Eurasianism and deride democracy while releasing a torrent of deception, disinformation, and de-truthing operations. Beijing has begun concentrating on "discourse power" as its way of influencing how people think about China and its growing reach around the world.

Meanwhile, some leaders of more-open societies have misconceived the concept of soft power, inflating it as *smart power*, and have neglected to come up with a doctrinal derivative that could rival hard power's realpolitik. Others have persisted with realpolitik, trying to adapt it to the information age. This ideational inertia, even complacency, has left the United States at a conceptual disadvantage that has turned into a strategic disadvantage. The U.S. conceptual arsenal is still sorely lacking in terms of understanding and applying soft power.

In short, adversaries have begun deploying aggressive soft-power strategies and tactics—lately called *sharp power*—far more adroitly than ever expected, catching Washington and other liberal capitals quite unawares and unprepared. And lately, rather than rethink matters, new leaders, especially in Washington, have continued to neglect soft-power capabilities and reverted to emphasizing hard power and realpolitik.

This state of affairs should be viewed with alarm; it should prompt an awareness of the urgent need to rethink statecraft for the information age. For us, this means shifting away from realpolitik toward noopolitik, a concept inspired by a second stream of thought.

Geosphere, Biosphere, and Noosphere

Over the past century, various scientists in Europe, the United States, and Russia have worked on developing a stream of thinking about the geosphere, biosphere, and noosphere. Whether appearing singly or jointly, these three terms should be viewed as a set for understanding Earth's evolution as a planet. First to evolve was a globe-circling *geosphere*, consisting of a geological mantle. Next was a globe-circling biological layer, or *biosphere*, consisting of plant and animal life. Third to grow and develop will be a globe-circling realm of the mind, a "thinking layer" termed the noosphere. These concepts were all in use by the 1920s, and they continue to spread today. Chapter Two discusses this in detail.

The term *noosphere* emerged when French theologian-paleontologist Pierre Teilhard de Chardin, French mathematician Édouard Le Roy, and visiting Russian geochemist Vladimir Vernadsky met in Paris in 1922 to speculate whether, because of humanity's growth, Earth would ultimately evolve a new layer: an all-enveloping noosphere (from the Greek word *noos*, meaning "the mind"). Teilhard defined the *noosphere* as a globe-circling "realm of the mind," a "thinking circuit"—in the later words of his colleague, Julian Huxley, a "web of living thought" and "a common pool of thought" that would lead to an "inter-thinking humanity" (Teilhard, 1965). For Teilhard, it was both a spiritual and a scientific concept; for Vernadsky, it was strictly a scientific concept—though both regarded it as having democratic political implications.

At first, the concept of a noosphere spread slowly and selectively among environmental scientists and social activists in the West. Some early believers are credited with helping inspire the creation of the United Nations; United Nations Educational, Scientific and Cultural Organization (UNESCO); and other "noospheric institutions" after World War II. The concept attracted wide attention in Europe and the United States in the 1950s and 1960s, following the posthumous publication of Teilhard's books *The Phenomenon of Man* and *The Future of Man*, which became bestsellers. Even so, the concept still spread mostly among individuals—until the 1990s.

Since then, the rise of the internet has excited a sense among myriad theorists and prophets of the information age that cyberspace is providing a technical foundation for the emergence of the noosphere. Thus, while the concept still has not gone mainstream, it is now proliferating far and wide at the level of online platforms, not just individuals; *Wired* magazine, the website Edge.org, the Evolution Institute, and various magazines and websites associated with pro-commons social theory and social activism often feature articles supporting the concept's potential.

Lately, various technologists and other scientists have preferred to develop newer concepts—e.g., collective consciousness, the global brain—that are not focused exactly on the noosphere but that still descend partly from it. Moreover, future successes with these alternate concepts are bound to help further the noosphere concept. It is here to stay, and it will continue growing in significance.

Noopolitik as an Approach to Statecraft

In sum, the noosphere concept provides solid logical grounding for thinking about policy and strategy in the information age. Hence, our derivative concept—noopolitik—matches up with soft power in the way that realpolitik matches up with hard power. No alternative concept does this as well; the noosphere offers the broadest way to think about information-based realms and their dynamics.

We first proposed noopolitik as an alternative to realpolitik back in 1999, but little happened to further its development. Since then, other strategists have proposed kindred concepts—e.g., *cyberpolitik, netpolitik, infopolitik, information engagement, information statecraft, information geopolitics*—and they, too, have failed to gain traction. Individually, the definitions of these kindred concepts might vary somewhat, but what is more important is that, collectively, all of these concepts represent innovative but so-far-unsuccessful efforts to improve the U.S. conceptual arsenal for dealing with information-age threats, challenges, and opportunities.

All of this leads to two points. First, noopolitik remains a suitable proposal for reorienting statecraft in the information age. Next, even if our particular concept does not take hold, strategists had better come up with something very similar—and quickly, before the world's dark adversaries do irreversible harm to the United States and other open societies by continuing to apply their own vexing mutations of noopolitik.

Taken seriously, the noosphere concept has particular implications for developing noopolitik as an approach to statecraft. The noosphere began as a scientific and spiritual concept, but it has acquired a forward-looking political cast. Its expansion implies the ascendance of ideational and other soft-power matters. The noosphere favors upholding ethical and ecumenical values that seek harmony and goodwill, freedom and justice, pluralism and democracy, and a collective spirit harmonized with individuality. It is an anti-war and pro-environment concept. Strategically, it implies thinking and acting in global or planetary ways while minding long-range ends and the creation of new modes of agency to shape matters at all levels. It implies humanity coming together through all sorts of cognitive, cultural, and other close encounters. It is about the coevolution of the planet and humanity—thus, it implies understanding the nature of social and cultural evolution far better than theorists have up to this point. And it means engaging nonstate, as well as state, actors in a quest to create a new (post-Westphalian) model of world order that is less tethered to the nation-state as the sole organizing principle and focus of loyalty. Furthermore, the noosphere favors the widespread positioning of sensory technologies and the creation of sensory organizations for planetary and humanitarian monitoring purposes.

Yet, positive and peaceful as all of this might seem, growth of the noosphere also implies having to deal with myriad ideational clashes and conflicts. Indeed, Teilhard, Le Roy, and Vernadsky said to expect ruthless struggles, shocks and tremors, and even an apocalypse as different parts of the noosphere begin to fuse around the world. Altogether, these are not implications that the founders simply tacked on; rather, they are drawn from observing principles and dynamics that attended the prior development of the geosphere and biosphere as global envelopes.

Proponents and practitioners of noopolitik should heed these distinctive implications and should not treat noopolitik as a self-aggrandizing public-relations or propaganda game. When the switch to noopolitik deepens in the decades ahead, strategists will gradually figure out how different it is from realpolitik, for noopolitik requires a fresh way of looking at the world—a new mindset, a different knowledge base, a different assessment methodology. How to look at hard power, and thus realpolitik, is quite standardized by now. But how best to understand and use soft power is far from settled. Noopolitik depends on knowing—and finding new ways of knowing—about ideational, cognitive, and cultural matters that have not figured strongly in traditional statecraft. It will eventually be seen that noopolitik is not only an information-age alternative to realpolitik but also a prospective evolutionary successor to it.

Because noopolitik is ultimately about whose story wins, narratives will have to be carefully crafted to suit the context. That narratives are crucial for maneuvering in today's world is widely accepted, but designing them remains more of an art than a science, and there is still plenty of room for new ideas about how to build and wield expertise. For example, the American effort to promote democracy abroad has proceeded unsuccessfully, even defectively, for many years. How best to promote democracy might well become a key challenge for noopolitik, and the answers might prove quite different from what has been assumed under past grand strategies.

The following are some of the steps that we recommend to help enable and energize a shift to noopolitik:

- Rethink *soft power*, especially its dark sides. We should not have to list this; it should be cleared up by now—but it is not.
- Create international "special media forces" that could be dispatched into conflict zones to help settle disputes through the discovery and dissemination of accurate narratives and for purposes of rumor control.

- Uphold "guarded openness" as a strategic principle: This means remaining open (particularly among allies) in accordance with democratic values while also creating mechanisms for guardedness to allay risks inherent in being open.
- Take up the cause of the "global commons" as a pivotal issue area: Though valued by many civilian activists and military strategists, this concept has yet to gain public recognition, and it is presently under challenge from arch-traditionalists who prefer a return to nationalist or neomercantilist policies in the name of sovereignty.
- In the case of the United States, institute a government requirement for periodic reviews of the nation's "information posture": One's information posture toward allies and adversaries is now as crucial as one's military posture. The latter receives regular review; it is time to figure out how best to assess and enhance national information postures as well.

Such measures, along with others mentioned in this Perspective, can open up transformational possibilities and opportunities for shifting from realpolitik to noopolitik as the basis of a new mode of statecraft attuned to the information age. They could help burnish the image of the United States and its allies in the world once again; lessen the bitterness and violence of conflicts; revitalize diplomacy, especially public diplomacy; and set the world on course toward sustainable peace and prosperity—for, whereas realpolitik treats international relations as intractably conflictual, the starting point for noopolitik is faith in upholding our common humanity and a belief that, in statecraft, ideas can matter more than armaments.

Even now, many shifts, risks, and conflicts that are commonly categorized as geopolitical in nature are, on closer examination, primarily noopolitical. For example, during the past decade, the Arab Spring in the Middle East, the rise of the far right in Europe, Hindi-Muslim clashes in South Asia, and protest movements in Hong Kong all have had geopolitical implications; but they might be better understood as essentially noopolitical in nature. Around the world, myriad cognitive wars—ideological, political, religious, and cultural wars—are underway, aimed at shaping people's minds and asserting control over this or that part of the emerging noosphere. At the same time, people are searching for new ways to get along and cooperate in addressing such global challenges as climate change and refugee settlement. Here, too, policies and strategies guided by noopolitik rather than realpolitik might well fare better for the common good of all parties.

Acknowledgments

For assistance in helping shape this think piece one way or another, the authors' thanks extend within RAND to Matt Byrd, Paul Davis, Saci Detamore, Agnes Schaefer, Tom Szayna, Rand Waltzman, and Emily Ward; and outside RAND to Michel Bauwens, David Brin, Scott Jasper, Richard O'Neill, Nancy Roberts, and Paul Saffo. The authors are also grateful for encouragement from Nancy Snow and Nicholas Cull, who published an early version of this paper in their recent volume, *Routledge Handbook of Public Diplomacy*, Second Edition (Routledge, 2020).

Abbreviations

AI	artificial intelligence
DARPA	Defense Advanced Research Projects Agency
DoD	U.S. Department of Defense
FFRDC	federally funded research and development center
GCP	Global Consciousness Project
GEF	Global Environment Facility
IE	The Evolution Institute
IGO	international governmental organization
IS	Islamic State
JAM-GC	Joint Concept for Access and Maneuver in the Global Commons
NATO	North Atlantic Treaty Organization
NGO	nongovernmental organization
P2P	peer-to-peer
UN	United Nations
UNEP	United Nations Environment Programme
UNESCO	United Nations Educational, Scientific and Cultural Organization

CHAPTER ONE

Looking Back, Looking Ahead: America's Conceptual Arsenal

A few decades ago, it was still easy to be optimistic about the positive promises of the digital information revolution, but no longer. The world today is awash in dark new forms of warfare that are based on the nature of information, broadly defined. Information itself has become a significant form of power, a force multiplier for both good and evil, a great disruptor and force divisor, and a catalyst for social cohesion. Strategists of all stripes—theorists and practitioners—remain unsettled and often baffled about how best to analyze, organize, and act amid this stormy flux. All of the auguries suggest that matters could grow worse before they turn better in the years ahead.

In 1999, sensing early signs of this, we proposed that American statecraft adapt to the emerging information age by developing a new approach that we called *noopolitik*[1] (Arquilla and Ronfeldt, 1999). We derived this term from the Greek word *noös*, meaning "mind," drawing inspiration from the idea that a *noosphere*—a globe-circling "realm of the mind," indeed the grandest such realm—is emerging around the world. Thus, we urged strategists to think anew about the evolving nature of *information* and grasp that the worldwide growth and spread of the noosphere would profoundly affect statecraft in the decades ahead.

We further argued in 1999 that, as the information age deepened, noopolitik would prove a more meaningful approach for American statecraft and grand strategy than the prevailing approach commonly known as *realpolitik*—an idea coined in the mid-19th century in Germany that came to dominate statecraft and grand strategy throughout the 20th century. For practitioners of realpolitik, what matters are practical material realities that enable the use of hard power, such as military might, industrial strength, financial capacity, and natural resources.

---

[1] An alternative spelling might be *noöpolitik*, or even *noospolitik* or *noöspolitik*. However, all words in standard dictionaries that draw on the Greek word *noos* or *noös* drop the letter *s* and the umlaut—e.g., *noocracy*, *noogenesis*, *nootropic*, of course *noosphere*, and even such emerging terms as *noopolitical* and *noological*. For this reason, we have preferred *noopolitik* (nü-oh-poh-li-teek). We would also note that the Greek root *noos* is sometimes spelled *nous*. Indeed, the word *nous* is in the dictionary and is sometimes used, especially in British English, as a word that refers to a person's mind or intellect—which raises the possibility of coining *noupolitik* or *nouspolitik*, but these spellings seem less attractive than *noopolitik*.

Realpolitik is essentially an industrial-era concept. The digital information age would, we posited, undermine the conditions for traditional strategies based on real-politik and its innate preference for material "hard power" and lead to new strategies based on noopolitik and its innate preference for ideational "soft power." In our view, realpolitik has been to hard power what noopolitik should be to soft power—a strategic approach attuned to changing conditions. Accordingly, we posited in 1999 that a rethinking would be needed because the decisive factor in the new global wars of ideas will increasingly become "whose story wins"—the essence of noopolitik—more than whose weaponry wins.[2]

What accounts for this shift? Both the noosphere and noopolitik concepts reflect a theme that has figured prominently in our work on the information revolution: the rise of network forms of organization that strengthen civil-society actors. Few state or market actors, by themselves, seem likely to have much interest in fostering the construction of a global noosphere, except in limited areas having to do with international law or political and economic ideology. The impetus for a global noosphere is more likely to emanate from nongovernmental organizations (NGOs), other civil-society actors (churches, schools), international governmental organizations (IGOs, such as the United Nations [UN]), and individuals dedicated to freedom of information and communications and to the spread of ethical values and norms. We believed in 1999 that it was time for state actors to begin moving in this direction, too, for power in the information age was going to stem, more than ever, from the abilities of state and market actors to work conjointly with networked civil-society actors.

Against this backdrop, we provided an update in 2007 on the promise of noopolitik (Ronfeldt and Arquilla, 2007) for the first edition of a handbook on public diplomacy (Snow and Taylor, 2009). In it, we summarized our 1999 report and added four new points that we discuss anew in later chapters in this piece:

- Other information-age concepts similar to noopolitik—e.g., netpolitik, cyber-politik, infopolitik—had appeared, but all (including noopolitik) were having difficulty gaining traction (as we discuss in Chapter Four).
- Even so, the concept of soft power continued to dominate strategic discourse in many government, military, and think tank circles, but its definition was flawed and lacked operational and normative clarity.
- Meanwhile, in nonstate arenas where noosphere-building ideas were taking hold, activist NGOs representing global civil society were becoming practitioners of noopolitik—but the most-effective practitioners arose in the realm of "uncivil society" where militant jihadis organized in global networks and outfitted themselves with sophisticated media technologies.

---

[2]   Because this phrase—"whose story wins"—has often appeared in other people's writings, usually without attribution or citation, perhaps we should clarify that we originated it while writing our first report about noo-politik (see Arquilla and Ronfeldt, 1999, p. 53)—hence its appearance as the title of this piece.

- Thus, we still argued that American diplomacy would benefit from a course correction in the direction of noopolitik. But we also cautioned that conditions for doing so were much less favorable than they had been when we had first fielded the concept a decade earlier—and propitious conditions seemed unlikely to reemerge anytime soon.

Today, noopolitik remains a promising concept for American and other liberal societies' information strategies. However, it is not alive and well in the United States, where even soft power is lately in decline as a strategic concept. Instead, U.S. adversaries are the ones who are working on developing noopolitik—but in dark ways and by other names—and they are using it against the world's democracies. These new circumstances might mean, to echo Charles Dickens (1859), that we are now living in "the worst of times"—yet, because of this adversity, potentially also in "the best of times"—for revisiting and embracing the promise of the noosphere and noopolitik.

Our initial writings analyzed at length the new importance of information and the nature and growth of three nested information-based realms: cyberspace, the infosphere, and the noosphere. Of the three, cyberspace is the smallest: the infosphere is much larger, for it includes both nondigital and digital systems, stocks, and flows of information; and the noosphere engulfs them both, partly because it is generally considered to be less of a technological realm and more of an ideational, cultural, and cognitive realm than the other two (Arquilla and Ronfeldt, 1999, Ch. 2, especially pp. 9–20). We provided this analysis back then to recommend that strategists begin to gravitate toward the noosphere concept. However, by now, the importance of information and those three realms is conceptually much more familiar to strategists.

Therefore, in this Perspective, we dive straight into discussing the noosphere concept in more detail—from its origins in the 1920s to the spread of its influence today, a century later. We do so despite knowing that this concept still has not gained widespread traction among information-age strategists, some of whom might still prefer the infosphere as a concept for framing their analyses of information-age security threats and societal challenges (e.g., Mazarr, Bauer, et al., 2019; Mazarr, Casey, et al., 2019). Yet, for our work, the noosphere concept remains more appealing and pertinent, partly because, as we lay out in the next three chapters, it is an expressly ideational and evolutionary concept that has inherently dynamic implications for statecraft and grand strategy. The same cannot be said for the cyberspace and infosphere concepts.

Compared with our 1999 and 2007 writings, then, this 2020 update offers fresh insights about both the noosphere and noopolitik—notably,

- an expanded discussion of Teilhard's, Vernadsky's, and Le Roy's foundational ideas about the noosphere (Chapter Two)
- a broadened report about the worldwide spread of the noosphere idea in recent decades, including in Russia (Chapter Three)

- a new assessment of the noosphere's strategic implications for noopolitik (Chapter Four)
- An updated discussion of state and nonstate uses of realpolitik and noopolitik (Chapter Five)
- a warning that Beijing, Moscow, and other state and nonstate actors are using dark forms of noopolitik against America, its allies, and its friends, while Washington devalues soft power and tries out "deal power" (Chapter Six)
- a first-time analysis arguing that the "global commons" might be a pivotal issue area affecting the prospects for the noosphere and noopolitik (Chapter Seven)

We proceed this way partly because we have learned more about the noosphere concept, and also because we have found new implications for noopolitik. To conclude this fresh assessment, we offer new recommendations for policy and strategy in light of the evolving strategic situations of the United States and its friends and allies (Chapter Eight).

We focus mainly on the United States. Our nation's long history of trying to spread peaceful, democratic ideals, taking public diplomacy seriously, and pioneering the concept of soft power means that Washington should be highly attuned and attracted to fostering the noosphere and applying noopolitik. That it lags rather than leads in the practice of this new mode of statecraft should be a source of urgent concern.

What we are trying to do here is offer ideas that might help U.S. strategists strengthen their "conceptual arsenal" for dealing with the information age. The first time this illuminating term was used in regard to statecraft, more than thirty years ago,[3] then–U.S. Secretary of Defense Caspar Weinberger was concerned about U.S. deterrence strategy and wrote,

> The world has changed profoundly since the 1950s and early 1960s, when most of our conceptual arsenal was formulated—so profoundly that some of these concepts are now obsolete. (Weinberger, 1986, p. 676)

Recently, Graham Allison has urged U.S. strategists to accept that spheres of influence will remain a central feature of geopolitics. He says that doing so might be difficult after so many heady decades of U.S. dominance:

> Yet it could also bring a wave of strategic creativity—an opportunity for nothing less than a fundamental rethinking of the conceptual arsenal of U.S. national security. (Allison, 2020, p. 40)

---

[3] The term *conceptual arsenal* is also associated with writings from decades earlier by French social theorist Pierre Bourdieu.

Weinberger and Allison, like most of their fellow strategists, believe primarily in hard power, as well as in the set of concepts that have grown around it over the past two centuries—e.g., realism, geopolitics, balance of power, realpolitik. A comparable conceptual arsenal has yet to develop around soft power. The occasional efforts to view idealism or liberal internationalism as parts of the soft-power bundle do not work well.

Today, this conceptual vacuum around soft power puts the United States at a strategic disadvantage. The concept of noopolitik can help remedy that, or so we hope.

CHAPTER TWO

Origins and Attributes of the Noosphere Concept

The growth of the noosphere explains why we opted for the concept of noopolitik, for the noosphere offers the broadest way to think about information-based realms. This term was coined by French theologian-paleontologist Pierre Teilhard de Chardin, French mathematician Édouard Le Roy, and visiting Russian geochemist Vladimir Vernadsky when they met together in Paris in 1922.

Our earlier writings credited only Teilhard with originating the noosphere concept. For this update, we add new findings about Vernadsky's and Le Roy's contributions.

Teilhard's Thinking About the Noosphere

In Teilhard's view—notably in *The Phenomenon of Man* (1965) and *The Future of Man* (1964)—people were beginning to communicate on global scales; thus, parts of the noosphere were already emerging. He described it variously in those books as a globe-circling "realm of the mind," as well as a "thinking circuit," "a new layer, the 'thinking layer,'" a "stupendous thinking machine," a "thinking envelope," a "planetary mind," and a "consciousness," where Earth "finds its soul." According to a metaphor that Teilhard favored about "grains of thought,"

> The idea is that of the earth not only becoming covered by myriads of grains of thought but becoming enclosed in a single thinking envelope so as to form, functionally, no more than a single vast grain of thought on the sidereal scale, the plurality of individual reflections grouping themselves together and reinforcing one another in the act of a single unanimous reflection. (Teilhard, 1965, pp. 251–252)

In the introduction of Teilhard's 1965 book, Julian Huxley further defined Teilhard's concept as a "web of living thought" and a "common pool of thought" (Teilhard, 1965, pp. 18, 20). He also praised Teilhard for coming up with "a threefold synthesis—of the material and physical world with the world of mind and spirit; of the past with the future; and of variety with unity" (p. 11). Huxley even urged that "we should consider inter-thinking humanity as a new type of organism, whose destiny it is to realise new possibilities for evolving life on this planet" (p. 20).

According to Teilhard, forces of the mind had been producing "grains of thought" and other pieces of the noosphere for ages. Increases in social complexity were also laying the groundwork for the noosphere's emergence. Thus, the noosphere was on the verge of achieving a global presence—its varied "compartments" and "cultural units" beginning to fuse. As Teilhard put it, equating cultures with species, "cultural units are for the noosphere the mere equivalent and the true successors of zoological species in the biosphere." Once a synthesis occurs, peoples of different nations, races, and cultures will experience "unimaginable degrees of organised complexity and of reflexive consciousness"—a planetary "mono-culturation" will take shape, yet somehow without people losing their personal identity and individuality (Samson and Pitt, 1999, pp. 76–79).

Fully realized, the noosphere will raise mankind to a higher evolutionary plane, even an "Omega point," shaped by a collective coordination of psychosocial and spiritual energies and by a devotion to moral, ethical, religious, juridical, and aesthetic principles. However, Teilhard counseled, "No one would dare to picture to himself what the noösphere will be like in its final guise" (Teilhard, 1965, p. 273). Indeed, the transition might not be smooth—a "paroxysm," a global tremor, or possibly an apocalypse might mark the final fusion of the noosphere (pp. 287–290).

Although Teilhard's noosphere concept is deeply spiritual and far less technological than cyberspace or the infosphere, he identified increased communications as a catalyst. Nothing like the internet existed in his time. Yet he sensed in the 1960s that 1950s-era radio and television systems were already starting to "link us all in a sort of 'etherized' universal consciousness" and that someday "astonishing electronic computers" would provide mankind with new tools for thinking (Teilhard, 1964, p. 162). Today, decades later, he is occasionally credited with anticipating the internet, as well as the recent trendy notion that Earth is transitioning from the Pleistocene to the Anthropocene age because human activity has itself become a kind of geological force.

Vernadsky's Thinking About the Noosphere

Vladimir Vernadsky likewise reasoned that Earth first evolved a geosphere and then a biosphere and that a noosphere would be next. Indeed, he wrote the first book titled *The Biosphere* (in 1926), treating life's planetary spread as a new kind of geological force. But although his views parallel Teilhard's, they also differ—Vernadsky's are much more materialist, in spots more mystical, and always less spiritual (he was an atheist).

According to Vernadsky's landmark paper—*The Transition from the Biosphere to the Noösphere*—a series of inventions (e.g., fire-making, then agriculture, and now modern communications technologies) had been generating "biogeophysical energy" (Vernadsky, 1938). This energy had enabled the development of the mind and its capacity for scientific reasoning and would lead "ultimately to the transformation of the biosphere into the noösphere" (p. 20). In other words,

> This new form of biogeochemical energy, which might be called the energy of human culture or cultural biogeochemical energy, is that form of biogeochemical energy, which creates at the present time the noösphere. (p. 18)

In this seminal 1938 write-up, Vernadsky further argued that the noosphere's creation has "proceeded apace, ever increasing in tempo" over the "last five to seven thousand years" despite "interruptions continually diminishing in duration" (p. 29). Eventually, prolonged growth should bring about "the unity of the noösphere," as well as "a just distribution of wealth associated with a consciousness of the unity and equality of all peoples" (p. 31). But although Vernadsky averred that it is "not possible to reverse this process," he expected that "the transitional stage" would involve "ruthless struggle" and "intense struggles" that might span several generations. Nonetheless, it seemed unlikely that "there will be any protracted interruptions in the ongoing process of the transition from the biosphere to the noösphere" (p. 30).

Vernadsky believed that this analysis was thoroughly scientific. Yet he still wondered whether it "transcends the bounds of logic" and whether "we are entering into a realm still not fully grasped by science" (p. 31). He even made positive closing references to Hindu philosophy and to the role of art in man's thinking.

Years later, despite his dismay about World War II's destructiveness, Vernadsky continued to look ahead with visionary optimism, still associating the noosphere's rise with the values of freedom, justice, and democracy. In "The Biosphere and the Noosphere," an article compiled from his earlier writings that appeared in the journal *American Scientist* in 1945, he observed:

> The historical process is being radically changed under our very eyes. For the first time in the history of mankind the interests of the masses on the one hand, and the free thought of individuals on the other, determine the course of life of mankind and provide standards for men's ideas of justice. Mankind taken as a whole is becoming a mighty geological force. There arises the problem of the *reconstruction of the biosphere in the interests of freely thinking humanity as a single totality*. This new state of the biosphere, which we approach without our noticing it, is the noösphere. . . .
>
> Now we live in the period of a new geological evolutionary change in the biosphere. We are entering the noösphere. This new elemental geological process is taking place at a stormy time, in the epoch of a destructive world war. But the important fact is that our democratic ideals are in tune with the elemental geological processes, with the laws of nature, and with the noösphere. Therefore we may

face the future with confidence. It is in our hands. We will not let it go. (Samson and Pitt, 1999, p. 99)

Throughout his varied writings about the noosphere, Vernadsky extolled the emergence of reason as a powerful, even geological force tied to the development of science and scientific thinking. He mostly viewed the noosphere as the "sphere of reason," the "realm of reason," the "reign of reason," and even "life's domain ruled by reason" (Vernadsky, 1997, passim).

Vernadsky's audience consisted mostly of fellow scientists in Soviet Russia, not policymakers. But he did occasionally argue that government administrators should attend to his findings and that "[s]tatesmen should be aware of the present elemental process of transition of the biosphere into the noosphere" (Samson and Pitt, 1999, p. 38)—a point we would reiterate on behalf of noopolitik.

Teilhard and Vernadsky Compared

Teilhard and Vernadsky shared a deep belief in Earth's evolutionary path: first, a geosphere; then, a biosphere; and next, a noosphere. Yet their views differed regarding both causes and consequences. Teilhard's views were more spiritually grounded than Vernadsky's; the latter argued that geological and technological forces could explain the noosphere's emergence. Yet, like Teilhard, Vernadsky expected the noosphere to have wonderful ethical consequences—"a just distribution of wealth" and "the unity and equality of all peoples" (Vernadsky, 1938, p. 30). And, although both viewed the noosphere as a realm of collective consciousness, neither saw it becoming a realm of uniformity. Both men valued individualism and variety. Both favored a future built on democracy. In ways that seemed contrary to Darwinian theory at the time, both also thought that evolution depended on cooperation as much as competition.

Both were quite unclear regarding what the transition to the noosphere would be like. They both made the transitional phase seem inevitable. Teilhard even made it seem alluringly smooth and peaceful—for the most part. Yet if the men had offered comparisons (which neither did) to the transitions to the geosphere and biosphere, they surely would have noted that evolution of any kind is often far from smooth and peaceful; indeed, it is often chaotic, disjointed, and violent. Fortunately, Teilhard and Vernadsky at least allude to this prospect—Teilhard by noting that a global tremor, even an apocalypse, might befall the final fusion of the noosphere; Vernadsky by noting the likelihood of intense ruthless struggles spanning several generations. Both recognized humanity's capacity for self-destruction.

Teilhard and Vernadsky were also unclear about another matter regarding the transition: Both saw the noosphere as evolving piecemeal around the planet, much as did the geosphere and biosphere, with some parts arising here and then spreading there, other parts elsewhere, with interconnections and interactions increasing over time, until the entire planet would be caught up in webs of creation and fusion. But neither Teilhard nor Vernadsky specified exactly what parts and pieces might matter along the way. Teilhard at least indicated that "compartments" and "cultural units" bearing "grains of thought" would do the "fusing." That is not much to go on, but it is helpful for thinking strategically, as we argue later.

Le Roy's Depiction of the Transition

Le Roy's few writings offer further insight into how the biosphere-noosphere transition might occur. In his 1928 volume, *The Origins of Humanity and the Evolution of Mind*, Le Roy offered a "hydro-dynamical" metaphor to depict how the noosphere might emerge. It would not be like the growth of a branching tree, but rather would occur in ways resembling watery spurts, jets, and spouts that eventually link together to form a layer that covers all of Earth. According to Le Roy,

> This is the noosphere, spurting and emanating from the biosphere, and finishing by having the same amplitude and same importance as its generator. . . . [It is] the spurting points that [will] attach the noosphere to the biosphere. (Samson and Pitt, 1999, pp. 66–67)

Metaphors aside, Le Roy went on to identify real-world factors that would drive creation of the noosphere: "division of work, game of association and habit, culture and training, exercise of all types; from where come social classes, types of mind, forms of activity, new powers" (Samson and Pitt, 1999, p. 67). This would finally result in a spiritualized separation of the noosphere—"a disengagement of consciousness . . . and the constitution of a superior order of existence . . . where the noosphere would strain to detach itself from the biosphere as a butterfly sheds its cocoon." A "mysterious force of thought cohesion between individuals" would provide the impetus for the layer's formation (p. 69).

Thus, much like Teilhard and Vernadsky, Le Roy viewed the expansion of the mind and the creation of the noosphere as a planetary process culminating in the noosphere's separation from the biosphere, though not without risks:

> We are, in truth, confronting a phenomenon of planetary, perhaps cosmic, importance. This new force is human intelligence; the reflexive will of humankind. Through human action, the noosphere disengages itself, little by little, from the biosphere and becomes more and more independent, and all this with rapid acceleration and an amplification of effects which continue to grow. Correlatively however, by a sort of return shock, hominisation has introduced, in the course of life, some formidable risks. (Samson and Pitt, 1999, p. 5)

This depiction of the noosphere's emergence is quite dramatic, even gripping and visionary—and so are the depictions offered by Teilhard and Vernadsky. Their metaphoric power might well help explain why the noosphere concept keeps spreading, not only over time but also across spiritual, intellectual, and other boundaries.

CHAPTER THREE

Propagation of the Noosphere Concept in Recent Decades

A proto-noosphere is already taking shape, albeit slowly, inchoately, and disjointedly. The more it matures and attracts attention, the more noopolitik will become applicable. As goes the noosphere, so goes noopolitik. Therefore, it behooves us to show how people's sense of the noosphere is spreading before we turn to updating the prospects for noopolitik—after all, both concepts are still unfamiliar to many people. This chapter confirms that the noosphere concept is increasingly taking hold, far and wide.

The concept spread initially in Europe and the United States after Teilhard's posthumous publications in the 1950s and 1960s, and in Russia after Vernadsky's return there in the 1920s–1930s. Good histories are yet to be written, but testimony to the concept's spread from the 1920s to the 1990s is documented in the exemplary compilation by Paul R. Samson and David Pitt, *The Biosphere and Noosphere Reader: Global Environment, Society and Change*. As Samson and Pitt show,

> The noosphere concept captures a number of key contemporary issues—social evolution, global ecology, Gaia, deep ecology and global environmental change—contributing to ongoing debates concerning the implications of emerging technologies such as human-created biospheres and the Internet. (Samson and Pitt, 1999, back cover)

This landmark volume presents writings from myriad philosophers, theorists, and scientists besides Teilhard, Vernadsky, and Le Roy—specifically, Henri Bergson, Julian Huxley, Arnold Toynbee, James Lovelock, Lynn Margulis, Rafal Serafin, Marshall McLuhan, Theodosius Dobzhansky, Dorion Sagan, Richard Dawkins, Kenneth Boulding, and Nikita Moiseyev, among others. In addition, former president of the Soviet Union Mikhail Gorbachev, another believer in the noosphere concept, wrote the book's foreword. The book features quite a set of luminaries.

Through their writings, the noosphere concept spread much as its originators had envisioned—as "grains of thought" beginning to fuse into a "common pool of thought" (Teilhard), and as "spurts," "jets," and "spouts" that may coalesce into a "layer" (Le Roy), eventually amounting to a veritable "geological force" (Vernadsky). This process got well underway as all of these proponents voiced their views as individuals, and then even more so when some deliberately inspired the creation of "noo-spheric institutions"—notably, the UN and the United Nations Educational, Scientific and Cultural Organization (UNESCO) in 1945 and Green Cross International in 1993, not to mention numerous social-activist civil-society NGOs (Samson and Pitt, 1999, pp. 184–185). Since then, much more has happened.

An Expanding Presence—From Individuals to Platforms

When we introduced noopolitik in 1999, the noosphere idea was attracting ever more interest and adherents—but mostly at the level of scattered individuals. Today, the concept still has not gone mainstream. Yet recognition now extends well beyond individuals; the concept has gained traction on many organizational platforms, especially those that have developed an internet presence—e.g., websites for magazines, institutes, blogs, and activist NGOs.

Peace and environmental movements were among the first to seize on the noosphere concept. Of special note, scholar-activist Elise Boulding, 1988, outlined, à la Teilhard, a "many-layered map of the world" that consisted of the geosphere, biosphere, and "sociosphere" (i.e., families, communities, nation-states, international organizations, and "the peoples' layer" of NGOs). At the top, she placed the noosphere, which consisted of "the sum total of all the thoughts generated in the sociosphere." Indeed, "[t]he more we can involve ourselves in the networks that give us access to that envelope, the more we can contribute to the emergence of that [global civic] culture" (pp. 54–55). Her writings, more than others at the time, showed that the noosphere concept was gradually gaining resonance and credibility among transnational civil-society actors, far more than among government and commercial actors.

As for new platforms, the sensational growth of *Wired* magazine, founded in 1993, greatly boosted recognition of the noosphere concept. In a renowned article, cyberspace guru John Perry Barlow observed in 1998 that

> The point of all evolution to this stage is to create a collective organism of mind. With cyberspace, we are essentially hardwiring the noosphere. (Kreisberg, 1995)

More to the point, Kevin Kelly, *Wired's* cofounder, first executive editor, and current senior maverick, writing independently in 2009, predicted an "emerging global superorganism" he called the *technium*—"the area where we have maximum machine connection and maximum human connection"—and he said it will engulf the noosphere:

> The entity represented by this space is not just the One Machine composed of all other machines connected together. Nor is it the Noosphere of all human minds connected into one large supermind. Rather it is the vigorous hybrid of both all

human minds and all artificial minds linked together. It is the pan-mind. At this juncture the nodes are anything that generates a signal—either humans or machines. (Kelly, 2009)

Kelly foresees this superorganism spreading to become a "planetary layer" that he calls the *holos*—again, as an all-engulfing variant on the noosphere:

A hundred years ago H. G. Wells imagined this large thing as the world brain. Teilhard de Chardin named it the noosphere, the sphere of thought. Some call it a global mind, others liken it to a global superorganism since it includes billions of manufactured silicon neurons. For simple convenience and to keep it short, I'm calling this planetary layer the holos. By holos I include the collective intelligence of all humans combined with the collective behavior of all machines, plus the intelligence of nature, plus whatever behavior emerges from this whole. This whole equals holos. (Kelly, 2016, p. 192)

Another luminous new platform is the website Edge.org, launched in 1996 under the leadership of John Brockman as a gathering place for a rich variety of leading "third culture" thinkers from around the world. Each year, they are consulted to compile answers to Edge.org's renowned Annual Question. Their answers sometimes include the noosphere. Thus, in response to the 2010 Annual Question—"How Is the Internet Changing the Way You Think?"—psychologist Mihaly Csikszentmihalyi replied,

The development of cooperative sites ranging from Wikipedia to open-source software (and including *Edge?*) makes the thought process more public, more interactive, more transpersonal, resulting in something similar to what Teilhard de Chardin anticipated over half a century ago as the "Noösphere," or a global consciousness that he saw as the next step in human evolution. (Csikszentmihalyi, 2011)

And to the 2017 Annual Question—"What Scientific Term or Concept Ought to Be More Widely Known?"—historian David Christian replied, "The Noösphere":

The idea of the "Noösphere," or "the sphere of mind," emerged early in the 20th century. It flourished for a while, then vanished. It deserves a second chance. . . . Freed of the taint of vitalism, the idea of a Noösphere can help us get a better grip on the Anthropocene world of today. (Christian, 2017)

Yet another platform, the Evolution Institute (IE), a nonprofit think tank founded in 2010, aims to apply evolutionary science to social issues—spirituality included. Its activities often bring up the noosphere. An early IE conference held that the growth of the noosphere helped explain "why the current intersection of science and spirituality explored in this roundtable is so exciting and qualifies as a genuinely new synthesis" (Wilson et al., 2015). The odd mix of evolutionary scientists and spiritual visionaries at the conference also discussed the possibilities of "conscious evolution"—a rather Teil-hardian notion. As a follow-up, in 2019, IE produced a long series of online posts about "Conscious Evolution" in which the noosphere often figures.

IE's leader, evolutionary biologist David Sloan Wilson, often praises the concept in his own writings. His book *The Neighborhood Project: Using Evolution to Improve My City, One Block at a Time* (2011) has a chapter titled "We Are Now Entering the Noosphere" that equates small groups and their activities with Teilhard's "grains of thought" (see also Tippett, 2014; Wilson, 2012). Wilson's latest book, *This View of Life: Completing the Darwinian Revolution*, speculates further about "grains of thought" coalescing into a noosphere—in his view, à la Teilhard, "they will eventually merge with each other to form a global consciousness and self-regulating superorganism called the Omega point" (Wilson, 2019, pp. xi–xiv).

In yet another significant development, various online platforms have appeared, all on the Left,[1] that favor peer-to-peer (P2P) and pro-commons social theory and social activism—e.g., Commons Transition, Integral World, *Kosmos Journal*, On the Commons, the P2P Foundation blog, the P2P Foundation Wiki, Shareable. Those who post on these platforms often comment favorably about the noosphere concept, thereby aiding its spread.

Here are two good examples: Eminent *Kosmos* contributor James Quilligan included the noosphere in his layout of "the global commons" by proposing the following:

> Today's global superbubble is the result of deep structural imbalances between economic ideology and policy (noosphere), and environment and labor (biosphere) and physical resources (physiosphere). The challenge is to assemble international representatives from all regions and sectors to discuss global commons issues in a negotiating format which integrates these three streams of evolution. (Quilligan, 2010)

And at *Commons Transition*, leading P2P theorist Michel Bauwens made a similar point during a long interview about future social transitions (we appreciate his nod to noopolitik):

> Right now, the nation-state is no longer a key instrument of change, so we must focus on building transnational open source communities of collective intelli-gence, i.e. a noopolitik for the noosphere. (Bauwens, 2018)

---

[1] Although people and platforms on the Left have shown interest in the noosphere concept, interest on the Right is rare. Theorists and activists on the Right are deeply interested in information-related concepts, sys-tems, technologies, and their effects—but they prefer such traditional concepts as "culture," "ideology," and "the media," maybe even "atmosphere" or "zeitgeist," over "noosphere" or other futuristic concepts noted in this chapter (see Mudde, 2019).

Medium magazine, founded in 2012, is another left-leaning publishing platform where the noosphere often receives attention. Particularly noteworthy is a recent article by Peter Limberg and Conor Barnes on "memetic tribes" and how they fight over culture-war issues in the noosphere:

> Memetic tribes are multitudinous, fractious, unscrupulously optimistic, and divide the world into allies and enemies. They are locked in a Darwinian zero-sum war for the narrative of the noosphere, the sphere of human thought. (Limberg and Barnes, 2018)

Indeed, the crises that have ignited this "Culture War 2.0" amount to "dynamite distributed throughout the noosphere." And the twenty-four "memetic tribes" participating in it correspond to "fragments of the larger noosphere." If countermeasures are not found, the authors "expect Culture War 2.0 to slink closer and closer to kinetic warfare, whether '60s style showdowns or more dramatic escalations" (Limberg and Barnes, 2018). This is quite a warning—and we share their concern.

A more questionable entry in this survey, but one that should not be omitted, is the unusual, controversial, inconclusive Global Consciousness Project (GCP). Psychologist Roger D. Nelson led it at Princeton University during 1998–2015, as

> an international collaboration of researchers interested in the possibility that we can detect faint glimmerings of a coalescing layer of intelligence for the earth, what Teilhard de Chardin called the Noosphere. (Nelson, undated)

Mostly a parapsychology experiment, the GCP deployed engineering devices around the world to try to detect whether a collective consciousness might be forming in response to major world events (e.g., the September 11, 2001, terrorist attacks on the United States). According to Nelson, undated,

> Suggestions like those made in many intellectual and cultural traditions, that there is an Earth consciousness, appear to have a modicum of scientific support in the GCP results . . . and that we may be interconnected on a grand scale by consciousness fields.

This is not exactly the sort of validation we are looking for, but it does represent another effort to investigate this "realm of the mind."

Lately, even scientists at the Defense Advanced Research Projects Agency (DARPA) have shown passing interest in the noosphere. Note the following agenda item for a recent conference:

> Noosphere: Create, measure, and model foundational questions regarding humans, human-machine interactions, and society[.] For example, are there new approaches to 'computation' based on human or animal social or cognitive processes and how

might we understand them? We are also discussing how human perception might be a tool in modern conflict resolution. (DARPA, 2017)

When we first wrote about these matters (Arquilla and Ronfeldt, 1999), this altogether diverse array of pro-noosphere platforms did not exist. And it was barely emerging at the time of our first update (Ronfeldt and Arquilla, 2007). Today, as the preceding paragraphs have shown, pro-noosphere platforms now have a fairly strong and growing presence, enabling many new voices to join in the ever-widening struggles over the future of the noosphere. (Besides, people have meanwhile adapted and then taken to such new terms as "blogosphere" and "Twittersphere"—can popular recognition of the noosphere be that far away?)

Meanwhile, in Russia

Far away, as a result of Vladimir Vernadsky's early work on the biosphere and noosphere, as well as work by fellow scientists (e.g., Alexey Eryomin, Nikita Moiseyev), *noos*-related concepts have grown in stature in Russia more than has been recognized. They reportedly continue to flourish in subgroups within the Russian Academy of Sciences, notably at the Vernadsky Institute of Geochemistry and Analytical Chemistry and the Institute for the Scientific Research and Investigation of Cosmic Anthropoecology.

Elsewhere, Russians also lead the Noosphere Spiritual Ecological World Assembly, which holds periodic conferences that attract New Age believers from around the world, notably José Argüelles, author of *Manifesto for the Noosphere: The Next Stage in the Evolution of Human Consciousness* (2011). Other spin-offs from Vernadsky's thinking include the Galactic Research Institute and its Foundation for the Law of Time, along with an online activity it organized in 2012, the First Noosphere World Forum. These (and other) New Age activities might not matter for thinking about American information strategy and diplomacy, but they do indicate the influences Vernadsky and his Russian scientist colleagues have had, not only in Russia but also in odd circuits around the world. In particular, the far-right radical Lyndon LaRouche became a devotee of Vernadsky's ideas, which are still promoted by his followers.[2]

Further extending Vernadsky's influence (and evoking the Global Consciousness Project at Princeton), Russian eclectic Anton Vaino coinvented and touted what he called the *nooscope* during 2011–2012 as "a device that records changes in the noosphere . . . the first device of its kind that allows for the study of humanity's collective mind" (Hartshorn, 2016; Stanley, 2016). If fully operationalized, it would deploy a complex system of "sensory networks" around the world to collect data and scan activities in seven areas he named: the business sphere, market conscience, the infra-

---

2   See Foundation for the Law of Time, undated; Noosphere Forum, 2020; and LaRouche PAC, undated.

structure of human life-support systems, technogeneous catastrophes, natural disasters, special purpose layers, and collective consciousness. Vaino's influence and the nooscope's purpose and status are unclear. But, curiously, Russian President Vladimir Putin appointed him chief of staff in 2016, a position he holds today. This has aroused speculations about whether Putin's ideas for a "Third Way" and "managed democracy" might mean that he aims to install a *noocracy*—a term from Plato, reiterated by Vernadsky, that means "rule of the wise"—for authoritarian mind-manipulating purposes.

Decades earlier, outside Russia, two of that country's stellar émigrés, Gorbachev and Andrei Sakharov, added to pro-noosphere thinking and activism in the West— Gorbachev through his opposition to nuclear war and support for humanitarian relief (Gorbachev, 1986), Sakharov through his concerns about the global commons (Goodby, 2015a; Goodby, 2015b).

Ruminations on Collective Consciousness

Throughout history, every expansion in interpersonal communications and connectivity has led to speculation that a collective, even global consciousness might be forming. The noosphere is but one of many concepts for grasping this. Key 19th-century precursors were Georg Wilhelm Friedrich Hegel's "objective Spirit," Ralph Waldo Emerson's "Over-Soul," and Emile Durkheim's "collective consciousness." The early 20th century brought Henri Bergson's "creative evolution," Carl Jung's "collective unconscious," and H. G. Wells's "world brain."

In the late 20th century, ideas multiplied that a collective intelligence, global consciousness, global brain, and/or global mind may someday arise from cyberspace and the internet. Some of these ideas reflect Marshall McLuhan's 1962 notion of the "global village," whereby

> Electric circuitry has overthrown the regime of 'time' and 'space' and pours upon us instantly and continuously concerns of all other men. It has reconstituted dialogue on a global scale. Its message is Total Change, ending psychic, social, economic, and political parochialism. . . . Ours is a brand-new world of allatonceness. 'Time' has ceased, 'space' has vanished. We now live in a global village . . . a simultaneous happening. (McLuhan, Fiore, and Agel, 1967, pp. 16, 63)

Some also reflect Benedict Anderson's 1991 concept of *imagined communities*. He initially applied it to the formation of nations, but it can be extended to refer to Christendom, Islam's *Umma*, and other grand constructions (e.g., China's call for building "a community of common destiny for mankind" [Mardell, 2017]) that depend on tribe-like fraternity and brotherhood but not necessarily on physical territory:

It is *imagined* because the members of even the smallest nation will never know most of their fellow-members, meet them, or even hear of them, yet in the minds of each lives the image of their communion. . . . In fact, all communities larger than primordial villages of face-to-face contact (and perhaps even these) are imagined. Communities are to be distinguished, not by their falsity/genuineness, but by the style in which they are imagined. (Anderson, 1991, p. 6)

Now, still relatively early in the 21st century, more such ideas have emerged, prompted by further advances on the internet (e.g., Google, Facebook, cell phone access) and even more by the burgeoning promises of automation, big data, and artificial intelligence. Some expansive ideas reflect dark hopes—as in Abu Mus'ab al-Suri's concept of a *virtual caliphate* (an entity located not on a piece of land, but on a network in cyberspace) ruling over the Islamic *Umma* around the world (Lia, 2009) or racist notions of an "Aryan nation" spanning continents. But most of the newer ideas are, like the noosphere concept itself, benignly generalized. They include Kevin Kelly's concepts of the *technium* and *holos*, as discussed earlier. In addition, the recent trendy concept of the *Anthropocene*—a looming geological age in which Earth's evolution is driven primarily by human rather than natural forces—is meant to raise people's consciousness about humanity's collective environmental effects around the world (Brannen, 2019; Sideris, 2016). Making matters more nebulous and mysterious, philosophers interested in consciousness in terms of quantum dynamics have lately proposed *panpsychism* and *cosmopsychism* (Goff, 2017; Goff, 2018) to claim that unseen quantum fields and forces might underlie human consciousness and potentially enable a collective consciousness that extends well beyond Durkheim's original formulation.[3]

Some ideas have turned into enormous research and development projects. Two prominent centers are the Massachusetts Institute of Technology Center for Collective Intelligence and the Global Brain Institute. Both loosely reflect the noosphere concept (Heylighen and Lenartowicz, 2017, p. 3) but are more intent on studying how to improve collective problem-solving, how internet- and web-based connectivity might shape the nature of future information societies, and how distributed systems for collective intelligence might give rise to collective consciousness or a global brain. These questions pertain to the noosphere concept as well, but, to our understanding, it is more concerned with the kinds of values, narratives, and storylines that matter most for humanity's evolution.

---

[3]   Also pertinent might be biologist Rupert Sheldrake's arguments that unseen "morphic fields" might exist, and that these might help explain mental telepathy and other psychic phenomena affecting consciousness (Horgan, 2014). In addition, numerous science-fiction novels have involved mechanisms for collective consciousness (e.g., Brin, 1990).

Coda: Long-Range Future Viability of the Noosphere Concept

In sum, we have learned much that is new since our update in 2007. A few points stand out:

- The noosphere concept is spreading far and wide. Although it has not yet gone mainstream, pro-noosphere platforms have developed a strong attractive presence, enabling many new voices to join in ever-widening discussions about the noosphere's future. Most of this activity is in Europe and the United States, especially among NGOs, but more is occurring in Russia than we previously knew about.
- Politically, people and platforms on the Left have shown the greatest interest in the noosphere and its future prospects. Interest on the Right is rare.
- Concepts not focused exactly on the noosphere—e.g., collective consciousness, global brain—still descend partly from it. Moreover, future successes with these alternative concepts are bound to help further the noosphere concept as well. It is here to stay; it will continue growing in significance.
- Assessing the current state of the noosphere remains a problematic task. Little has been accomplished in this regard; no systematic methodology exists. But this also is the case with other big-information concepts, such as the technium, collective consciousness, and the global brain. Even so, bits and pieces of progress are being made: a recent focus on memetic tribalism is one example; another is the enduring concern about the waxing and waning of universal values.

Compared with other concepts about information power, the noosphere provides the best grounding for thinking about policy and strategy in the information age. Our derivative concept—noopolitik—matches up with soft power in the way that realpolitik matches up with hard power. No alternative concept does this as well.

Next, we update the implications of all of this for noopolitik.

CHAPTER FOUR

From Rethinking the Noosphere to Rethinking Noopolitik

As the prior chapter confirmed, the noosphere concept keeps gaining ground across myriad fields of inquiry and activism. As a result, two questions that Samson and Pitt posed two decades ago in *The Biosphere and Noosphere Reader* (1999) make even more timely sense today, for public policy dialogue and strategy increasingly merit rethinking in light of the noosphere's growth:

> Once again, we are faced with two questions: in what direction does public opinion want the noosphere to go and in which directions is the noosphere capable of going? Practically speaking, and in today's world, this translates into asking how the noosphere can be applied to help to solve problems in such areas as environment, health, poverty, violence and inequality (Samson and Pitt, 1999, p. 181).

With a few word substitutions, those two questions pertain as well to the future prospects for the concept of noopolitik—our proposal for an approach to information strategy based on the rise of the noosphere: In what directions will security strategists want noopolitik to go? And in what directions is noopolitik capable of going? We turn to such questions in this chapter and the ensuing chapters.

Strategic Vectors That Span the Noosphere and Noopolitik

If and as noopolitik takes hold in the years ahead, we hope that its development will reflect a keen, clear grasp of the noosphere concept. The vision of noopolitik that we laid out in 1999 still reads well, even in light of our updated findings about the noosphere:

> In sum, noopolitik is an approach to diplomacy and strategy for the information age that emphasizes the shaping and sharing of ideas, values, norms, laws, and ethics through soft power. Noopolitik is guided more by a conviction that right makes for might, than the obverse. Both state and non-state actors may be guided by noopolitik; but rather than being state-centric, its strength may well stem from enabling state and non-state actors to work conjointly. The driving motivation of

noopolitik cannot be national interests defined in statist terms. National interests will still play a role, but should be defined more in society-wide than state-centric terms and be fused with broader, even global, interests in enhancing the transnationally networked "fabric" in which the players are embedded. While realpolitik tends to empower states, noopolitik will likely empower networks of state and non-state actors. Realpolitik pits one state against another, but noopolitik encourages states to cooperate in coalitions and other mutual frameworks. In all these respects, noopolitik contrasts with realpolitik (Arquilla and Ronfeldt, 1999; see also Ronfeldt and Arquilla, 2007).

However, our research for this update leads to more-precise points about the relationships between the noosphere and noopolitik. Vectors are embedded in the noosphere concept that should be carried over to help develop noopolitik:

- The noosphere was originally a scientific and spiritual concept. It arose from revolutions in thinking about science and evolution, about complexity and consciousness, about the world becoming ever more interconnected and interdependent, and about the importance of cooperation, as well as conflict and competition. The noosphere concept meant that the expansion of knowledge and reason was essential for humanity to attain its planetary potential and address matters that require holistic analyses and answers. It is a pro-science concept; by extension, noopolitik must be pro-science too.
- The noosphere has become a visionary political concept as well. But it is not utopian. It is an evolutionary *protopian* idea (see Kelly, 2011)—which means expecting "progress in an incremental way where every year it's better than the year before but not by very much" (Shermer, 2015). Accordingly, the noosphere concept implies that people should anticipate and shape what lies ahead, with a sense of grounded realism, as well as hopeful idealism. It implies living within the permissible limits of the biosphere, in part by attending to the effects of human activity, so that the biosphere and noosphere are kept in a mutually beneficial balance. Thus, the noosphere concept offers an engaging vision of the future. In a sense, then, a key purpose of noopolitik is to prepare the way for the age of the noosphere while also protecting the biosphere and geosphere.
- The noosphere concept favors particular value orientations—not as philosophical ideals that are tacked on but rather as patterns and dynamics that have attended the evolution of the geosphere and biosphere and that will prove best for protecting the geosphere and biosphere and creating the noosphere in the future. Thus, the concept's originators put the emphasis on values that are ethical and ecumenical, that seek harmony and mutual goodwill, freedom and justice, and pluralism and democracy. The concept calls for the world and its cultures to be open and inclusive in ways that foster both unity and variety, both a collective spirit and individuality—all to foster, in the words of Huxley, an "inter-thinking

humanity" (Teilhard, 1965, p. 20). This makes the noosphere a pro-humanity and antiwar concept, as well as a pro-environment concept. As Nikita Moiseyev once said, "entering the age of the noosphere requires the practical reconstruction of the worldwide order and the establishment of a new thinking, a new scale of values and a new morality" (Samson and Pitt, 1999, p. 171). Accordingly, then, noopolitik should seek to uphold the same kinds of values.

- Since the beginning, the noosphere's emergence has been a function of revolutionary advances in information and communications technologies. Some of the latest advances were identified in the previous chapter—e.g., the internet, artificial intelligence (AI). A point worth adding is that the noosphere's growth will also become a function of the development and distribution of all sorts of sensory apparatuses that will amount to what McLuhan, 1964, p. 1, called "extensions of ourselves." A revolution in networked sensory technologies is underway; its maturation might prove transformative for the noosphere's growth, perhaps especially for worldwide monitoring by NGOs. Noopolitik would benefit from advancing such technologies.

- The noosphere concept implies a set of standards for strategy. This is clearest if *strategy* is defined not only in traditional terms, as the art of relating ends, ways, and means, but also in cognitive terms, as an art of positioning for spatial, temporal, and agentic advantages—for valuing the noosphere strategically means thinking and acting in global/planetary ways (spatially) while minding long-range future end states (temporally) and creating new modes of action to shape matters at all scales (agentically). That is how noopolitik should be conducted.

- The noosphere concept has long implied a serious challenge to Westphalian realpolitik-type thinking that nation-states are the most important actors and that material factors matter most. As the information age deepens, other actors and factors will increasingly matter more. It is unclear what will supersede the Westphalian model—theorists raise many possibilities. But it is clear that noopolitik should not reinforce that old model; rather, it should work to find and shape what might advisedly become the next model. As noted earlier, working with IGOs and NGOs as "noospheric institutions" might be important for this; so might advancing the concept of the global commons (as discussed later). A general point for noopolitik is to build a world order that relies more on networks than hierarchies.

- The noosphere concept is about the coevolution of the planet and humanity. This means that noopolitik should be deliberately designed around an understanding of the dynamics of social evolution. The fact that there is no agreed-upon theory of social evolution does not obviate this concern. Grand strategies often rest on judgments about social evolution, and what strategists think (or dismiss) about social evolution might make a crucial difference. Indeed, ideas about social evolution have often shaped grand strategy: e.g., containment theory in the 1950s, modernization theory in the 1960s, democratic enlargement in the 1990s, plus "end-of-history," "clash-of-civilizations," and "export-of-democracy" ideas that affected strategic thinking during the global war on terrorism. Add up all of these ideas, and quite a case can be made that grand strategies often depend on what assumptions about social evolution are embedded in them. This might seem a passing matter in realpolitik, but it is a requisite concern in noopolitik. Sounder ideas about social and cultural evolution are needed for making U.S. grand strategy in the age of the noosphere.

Recent Alternatives and Complements to Noopolitik

We have not been alone in arguing that the information age would disturb grand strategy and diplomacy so thoroughly that new concepts would emerge. Accordingly, David J. Rothkopf, 1998, p. 326, proposed that "the *realpolitik* of the new era is *cyberpolitik*, in which the actors are no longer just states, and raw power can be countered or fortified by information power." David Bollier, 2003, p. 2, proposed *Netpolitik* as "a new style of diplomacy that seeks to exploit the powerful capabilities of the Internet to shape politics, culture, values, and personal identity." In Europe, Philip Fiske de Gouveia, 2005, pp. 8–9, offered *infopolitik* to name a new era for public diplomacy based on "proactive international communication" and "the projection of free and unbiased information." All of these innovative terms were posed in contrast to realpolitik, as information-age ideas for strategy and diplomacy that would emphasize networks more than hierarchies and nonstate actors more than state actors. Yet none referred to the noosphere, none took hold, and they all have fallen by the wayside. Nonetheless, they helped stir and spread a sense that something new is on the way—good reason for us to continue advancing noopolitik, as we did with our 2007 update.

Since then, other analysts have proposed additional such ideas. Discussions have arisen about reorganizing around the principle of "information engagement" as a way to improve information-sharing and policy coordination across military and diplomatic sectors, both inside and beyond the U.S. government. Lately, following the change in administrations, a new concept appeared in a 2017 U.S. national security strategy document: "information statecraft" (Trump, 2017, p. 35). In academia, a newer proposal recommends "information geopolitics" (Rosenbach and Mansted, 2019). All three of these proposals call for improving government-wide coordination and crafting better strategic narratives—useful advice—but the two latest seem quite defensive and reactive, and both tend to emphasize cyber dimensions above all else.

So far, these concepts have gained no significant traction within the U.S. government and little traction in other liberal societies. Nonetheless, they add to the fact that a search remains underway to adapt statecraft to the information age and find better ways to deal with information as a multifaceted new form of power—one that can be used to advance the interests of open societies but that can also be used against them.

We formulated noopolitik largely as a contrast to realpolitik, but it is worth noting that the concepts of geosphere, biosphere, and noosphere have led to interesting parallel concepts about geopolitics, biopolitics, and noopolitics. The terms *geopolitics* (also, *geopolitik*) and *biopolitics* were coined by Swedish political scientist Rudolf Kjellén in 1900 and 1905, respectively. The term *biopolitics* has also been promoted as a new form of power by French theorist Michel Foucault in his writings since 1975. The newest term, *noopolitics*—so close to *noopolitik*—was first employed by Italian sociologist Maurizio Lazzarato in 2004 and 2006 (Lazzarato, 2006) and developed thereafter by Italian theorist Tiziana Terranova (2007) and French essayist Idriss Aberkane (2015). All but Kjellén are prominent thought leaders on today's Left.

Usage of the terms *noopolitics* and *noopolitical* has appeared mainly among writers and audiences on the Left in Europe. But usage seems likely to spread, and not only because of our own writings. Of particular interest, noopolitics has attracted attention in Russia, as a strategy for manipulating the noosphere (e.g., Nikonov, Baichik, Puiy, et al., 2015; Nikonov, Baichik, Zaprudina, et al., 2015). Although most writings about noopolitics trace the origins of the term to our first report about noopolitik (Arquilla and Ronfeldt, 1999), a precise definition for *noopolitics* remains unsettled. Depending on the writer, the term refers to a communications-based way to shape people's thoughts and actions that goes beyond biopolitics (Terranova, 2007), to the "geopolitics of information" (Aberkane, 2015), and, as noted, to a strategy for manipulating the noosphere (Nikonov, Baichik, Puiy, et al., 2015).

Against this broad background, an even older concept keeps resurfacing as an alternative to realpolitik: namely, *idealpolitik* (see Goldmann, 2013; Krastev and Benardo, 2020; Kober, 1990; Wallensteen, 2002; Wallensteen, 2013). Its proponents see a contrast between realism and idealism as schools of thought and ways to approach strategy. They also recognize that realpolitik and idealpolitik often can go hand in hand, side by side. By implying that the world would be better off with less realpolitik and more idealpolitik, they stress the importance of ideals, morals, and ethics in defining national identity, purpose, and legitimacy. Indeed, they call for the primacy of moral principles over sheer power, and thus for steadfastly promoting freedom, democracy, and human rights.

Historically, this moralizing concept predates the information revolution and has little to convey about it or its implications, such as the rise of network forms of organization. In short, idealpolitik has some congruence with noopolitik, but it is not a suitable alternative to it. The two concepts should not be conflated—idealpolitik is extremely concerned with moralizing (e.g., see Power, 2019), is insufficiently future-oriented, and has no inherent connection to the information revolution and its implications.

This is not to say that noopolitik dismisses idealism. Instead, noopolitik addresses idealism in a different, broader, and more inclusive way. Traditional idealism seeks to maximize the application of one's own ideals and values. Not so with noopolitik—it turns that kind of idealism inside out, for noopolitik aims to recognize that all people, no matter whose side they are on, will have their own ideals and values and react accordingly. Noopolitik would attend to this in a contingently inclusive way, for it seeks not to impose but rather to encompass the roles that ideals and values may play for all parties—and thus to sense at the beginning whether particular measures might result, say, in harmonious attraction or vengeful blowback.

Noopolitik, unlike traditional realpolitik and idealpolitik, is inherently preoccupied with cognitive and evolutionary sensibilities. In a sense, noopolitik is an approach to strategy that seeks to maximize cognitive and evolutionary security—i.e., noospheric security. For example, consider U.S. strategies in Afghanistan and Iraq over the past several decades. In both zones, U.S. strategies largely overlooked the enduring strength of traditional tribal (and sectarian) ideals, values, and ways of doing things as preferred by local populations. In both zones, U.S. strategies were also presumptuous about U.S. capabilities to export and promote democracy into the local environments. Over time, then, these protracted efforts to reroute the currents of history and culture by armed force have failed, reflecting the increasingly limited effectiveness of hard-power politics, even when wedded to good intentions. Our sense is that strategies guided by noopolitik would have fared better for all parties, largely by taking better account of local cultural, cognitive, and evolutionary dynamics.[1]

It is unclear how all of this conceptual innovation, and wrangling, will work out.[2] Which terms will endure, and with what implications for theory and practice? Yet noopolitik and noopolitics are gaining increasing recognition. Moreover, there are increasing indications that information-age theorists and strategists in China and Russia, as well as in India, Iran, and France, to name a few, are concerned with their own and others' ideas about noopolitik and noopolitics.

Before long, noological forces might be deemed as significant as geological and biological forces, just as Vernadsky predicted. And strategists might attend to noopolitik and noopolitics as much as they have to realpolitik and geopolitics.

---

[1]   An interesting, well-intentioned effort to move in this direction appeared with the U.S. Army's establishment of the Human Terrain System in Afghanistan and Iraq, a field operation created on grounds that "Cultural knowledge and warfare are inextricably bound" (McFate, 2005, p. 42). But the operation and its teams foundered before long, for multiple design and bureaucratic reasons (McFate and Fondacaro, 2011).

[2]   Yet another *ism* at play in the field of international politics, but more academic in origin and consequence, is *constructivism* (see Finnemore, 1996; Wendt, 1999). Although we do not highlight it in this study, we recognize its ancillary significance. In contrast to realism, and partly in reaction to realism's emphasis on material forces, constructivism emphasizes the importance of ideational constructions and shared ideas in power politics. Thus, constructivism offers another hopeful sign that the changes we anticipate—the rise of the noosphere and noopolitik—might radically alter the nature of statecraft and grand strategy.

Coda

Because of the aforementioned, the emergence of the noosphere begs for strategic thinking, which is why we proposed noopolitik in the first place and why we are offering this update.

Not everyone might agree. We have seen arguments that cyberspace (a key component of the noosphere) is "ill-suited for grand strategic theories"—the challenges it poses and the technologies on which it depends are said to be changing too rapidly and uncertainly for grand strategic thinking, at least for now (Libicki, 2014, p. 33). Might such arguments be applied to the noosphere, as well? By comparison, the noosphere is an even more vast, more complex "space," and it is evolving just as uncertainly, if less rapidly. Also, the noosphere is more difficult to pin down than cyberspace.

Nonetheless, as this and ensuing chapters clarify, our view, along with other views included in this piece, is that the noosphere concept does indeed invite grand strategic thinking. For us, that means, above all else, continuing to advance the concept of noopolitik.

CHAPTER FIVE

Contrasting Paradigms: Evolving from Realpolitik to Noopolitik

Realpolitik has been to hard power what noopolitik could become to soft power. This would bring a revaluing and rebalancing of the strategic spectrum that runs from hard to soft power—a kind of paradigm inversion. Realpolitik often involves soft-power tactics in specific situations (e.g., propaganda, deception, psychological operations), but always as supportive adjuncts to strategy, not as core essences. Indeed, classical Realists, from Niccolò Machiavelli to the modern era, have tended to view "considerations of morality as irrelevant"; whereas "personal behavior is usually judged by an ethic of *intention* . . . that of the statesman is essentially one of *consequence*" (Stoessinger, 1965, p. 228, emphasis in original).[1] In contrast, noopolitik means that strategic narratives and the like, designed to have cognitive and other psychological effects on people's thoughts and actions, become the very essence of strategy. Airplanes, tanks, ships, troops, and other hard assets remain the defining "platforms" for realpolitik; by contrast, information systems, social media, and well-crafted storylines are becoming the defining platforms for noopolitik. Combined arms are a key concern in realpolitik; combined cognitions matter most in noopolitik.

There are many ways to contrast the two paradigms. Table 5.1 highlights the contrasts that we deem to be among the most important at present (see Bew, 2014, on the history of the realpolitik concept).

Although Table 5.1 provides a present-day contrast between realpolitik and noopolitik, we must stress that noopolitik is not the polar opposite of realpolitik. The polar opposite of realpolitik at present is idealpolitik, just as the opposite of realism is idealism (see Chapter Four). Noopolitik is a present-day alternative to realpolitik, but, in addition, it represents an evolutionary step beyond it. As mentioned earlier, realpolitik was originally an industrial-age concept; noopolitik is a postindustrial, information-age concept. Realpolitik is a concept from the past; noopolitik is a concept for the future.

---

[1] For a recent expression in much the same vein, see Walt, 2020.

Table 5.1
Contrast Between Realpolitik and Noopolitik

Realpolitik	Noopolitik
States as key unit of analysis	NGOs, networks included as key actors
Primacy of national self-interest	Primacy of shared interests
Primacy of hard power	Primacy of soft power
System is anarchic, highly conflictual	Harmony of interests, cooperation
Power politics as zero-sum game	Win-win as preferred game
Politics as unending quest for advantage	Politics as pursuing a *telos* (end purpose)
Alliances conditional (oriented to threat)	Alliance webs vital to security
Ethos is amoral, if not immoral	Ethics are crucially important
Behavior driven by interests, threat, power	Behavior driven by common values, goals
Balance of power as the "steady state"	Balance of responsibilities as the "steady state"
Power embedded in nation-states	Power embedded in "global fabric"
Guarded, manipulative about information	Devoted to information-sharing, inter-thinking

Fully developed, noopolitik could become as big an evolutionary step beyond realpolitik as realpolitik was to what preceded it in prior, preindustrial centuries. What came before realpolitik has never had a precise name, but the dynamics of statecraft, diplomacy, and strategy back then are easy enough to characterize. They often revolved around the reputations of kings, popes, aristocratic courts, powerful well-connected families, and the like—and even tribes and clans. In other words, they revolved around kinship dynamics, both bloodline and fictive kinship. At high aristocratic levels in particular, marriage diplomacy was a keen preoccupation, including for shaping military alliances, amassing wealth, and gaining new territory.

In short, to coin a term, those were ages when *kinpolitik* prevailed (Ronfeldt, 2005; Ronfeldt, 2007). Of course, there were harbingers back then of what eventually evolved into realpolitik—the Peloponnesian War might be the earliest exemplar, and Machiavelli an early champion. But it was not until the industrial age emerged two centuries ago that realpolitik rose fully into its own, thereby becoming as complex an evolutionary step beyond kinpolitik as noopolitik will be beyond realpolitik in the times ahead.

Information-age strategists will be tugged between realpolitik and noopolitik in the coming transitional decades. Although neither, in and of itself, amounts to grand strategy, each poses a very different way of approaching *all* strategy, but especially grand strategy. As noopolitik takes shape and gains adherents, it will serve sometimes as a supplement and/or complement to realpolitik and other times as a rival paradigm, competing for salience in policy and strategy settings. As the global noosphere expands over time, noopolitik will increasingly provide the more pertinent paradigm. Today, noopolitik might seem to resemble liberal internationalism, but the latter is an aging paradigm, aspects of which can and will be folded into noopolitik.

The potential rise of noopolitik seems bound to spark resistance from the devotees of realpolitik. In response, they might endeavor to broaden their favored concept so that it explicitly incorporates soft power, to such a degree that any successful information strategy can thus be called an expression of realpolitik, without reference to noopolitik. In our view, this would run counter to long-accepted definitions of realpolitik as a distinctive approach to power politics. Such a step would also create needless conceptual drift and confusion. Noopolitik differs sharply from realpolitik; noopolitik, not realpolitik, is soft power's natural home. This will become increasingly clear as we progress deeper into the information age.

Global Civil-Society Actors as Harbingers of Noopolitik

As we have observed since beginning our work on noopolitik, states will no doubt remain the international system's paramount actors for the foreseeable future. The nation-state remains valuable as an attractive focus of loyalty and an effective, resilient organizational form. The information revolution has led to changes in the nature of the state, but certainly not to its "withering away," as Karl Marx and Friedrich Engels so famously forecast (see Bloom, 1946). What will continue to happen is a transformation of the nation-state system, driven partly by the rise of networked nonstate actors who are growing in strength and influence. This has been the trend for decades, with business corporations and international regulatory regimes. The trend that is still building is a gradual worldwide strengthening of transnational NGOs that represent civil society, perhaps especially those with interests in building and managing global commons. As this occurs, there will be a further rebalancing of relations among state, market, and civil-society actors around the world—in ways likely to favor noopolitik over realpolitik.

Noopolitik upholds the importance of nonstate actors, especially from civil society, and depends on their playing strong roles. Why? NGOs (not to mention individuals) often serve as sources of ethical impulses (which is seldom the case with market actors), as agents for disseminating ideas rapidly, and as nodes in networked "sensory organizations" that can assist with conflict anticipation, prevention, and resolution. Indeed, because of the information revolution, advanced societies are on the threshold of developing vast sensory apparatuses for watching what is occurring around the world. This apparatus, viewed collectively, is not new, because it consists partly of established government intelligence agencies, corporate market research departments, news media, and opinion polling firms. What is new is the looming scope and scale of this sensory apparatus, as it increasingly includes networks of NGOs and individual activists who monitor and report on what they see in all sorts of issue areas. Older technologies that they used included open forums, specialized internet mailing lists, web postings, and fax-machine trees. Now, smartphones and social media are the principal tools for rapid dissemination. For example, using these tools to provide early warning about crises is a burgeoning area of attention and development among disaster-relief and humanitarian organizations. And when it comes to human-caused disasters, like the catastrophic civil war in Syria, the ability to make the world swiftly and widely aware of atrocities—such as those caused by chemical weapons attacks—reflects an increasing power to shape the global discourse and the responsive actions that might ensue.

Against this background, the states that emerge strongest in information-age terms—even if by traditional measures they might appear to be smaller, less powerful states—are likely to be the states that learn to work conjointly with the new generation of networked nonstate actors rooted in civil society. Strength may thus emanate less from the "state" per se than from the "system" as a whole. And this might mean placing a premium on state-society coordination, including the toleration of "citizen diplomacy" and the creation of "deep coalitions" between state and civil-society actors. In that sense, the information revolution is impelling a shift in emphasis from a state-centric to a network-centric world, which parallels an emerging shift in the military world from traditional "platform-centric" to emerging "network-centric" approaches to warfare, as well as to operations other than war. As for higher-order international security dynamics, the rise of noopolitik might also begin to displace the "arms race" paradigm, a function of weaponry, with an "organizational race" focused on building networks.

This should favor noopolitik. While realpolitik remains steadfastly imbued with notions of top-down control, noopolitik is more about "decontrol" as a way of letting go to allow self-organization (Arquilla and Ronfeldt, 1999, p. 21; Ronfeldt, 2009)—perhaps deliberate, monitored decontrol—so that state actors can better adapt to the emergence of independent nonstate actors and learn to work with them through networked mechanisms for communication and coordination. Realpolitik leans toward an essentially mercantilist approach to information, as it once did with commerce; noopolitik is neither mercantilist nor exclusionary by nature.

Networks of nonstate actors have a long record of influencing public discourse, not to mention national and sometimes international strategy and policy. Their power was perhaps first demonstrated in opposition to the slave trade in the first half of the 19th century (Hochschild, 2005), and their power proved broadly successful around the world in curtailing this practice—except in the United States, which nearly tore itself apart in a bloody civil war in response to the tensions sparked by the abolitionist movement. In the latter half of the 19th century, during the colonial "scramble for Africa," it was civil-society networks that helped rein in the brutal rule of King Leopold II of Belgium in the Congo (Hochschild, 1998; Pakenham, 1991). Elsewhere, another civil-society network succeeded in ending the gross humanitarian violations (including the creation of the world's first concentration camps) by the British government in South Africa and fostering a tolerant peace agreement that concluded the bloody Boer War (Bossenbroek, 2018).

The spirit of the 19th-century civil-society activists was revived in the great relief and aid programs formed in the wake of both world wars, and then in the anti–nuclear weapons movements that played a significant role in curbing the arms race and setting the stage for arms reductions (Knopf, 1998). More recently, civil-society networks have helped energize and mobilize international humanitarian assistance and disaster relief, from the catastrophic tsunami that struck Indonesia in 2004 to the latest responses to climate change–caused damage done to numerous Caribbean nations. Needless to say, the attention given today to the urgent need to address global warming is largely due to civil-society networks—energized at times by key individuals, such as Greta Thunberg, whose role with Fridays for Future is reminiscent of Jodi Miller's role in the 1990s International Campaign to Ban Landmines.

NGOs lack the resources to practice realpolitik; they are better suited to waging educational and reputational noopolitik. The digital information revolution has greatly invigorated their capabilities for doing so.

Early Glimmers of Noopolitik During the Cold War Years

On the path to the noosphere, Vernadsky predicted an indeterminately long transitional stage marked by "ruthless struggle." In our view, first expressed in 1999, and to which we still adhere, this struggle is largely framed by the persistence of traditional realpolitik—whose basic tenets treat the world system as anarchic and highly permissive of violence—and the rise of new kinds of network-driven interactions associated with the rise of the noosphere. What we call noopolitik is struggling to emerge against the headwinds of a world paradigm that has dominated policy and strategy since the days of Thucydides two-and-a-half millennia ago.

Early glimmers of noopolitik and the emergence of a transnationally networked fabric of a global society could be glimpsed 70 years ago, in the creation of the UN and UNESCO in 1945, and were followed by the creation of the Universal Declaration of Human Rights and the Convention on the Prevention and Punishment of the Crime of Genocide in 1948. Later, at the height of the Cold War, the 1975 Helsinki Accords reflected similar global values, as did the Biological Weapons Convention of the same year. In a similar spirit, the Chemical Weapons Convention took effect in 1997. All of these are far more redolent of noopolitik than of realpolitik. They were all state-centric, but NGOs played decisive background roles.

As for nuclear abolition, this matter is moving more slowly. Atomic weaponry is deeply rooted in realpolitik, as former U.S. Secretary of State Henry Kissinger argued in his classic *Nuclear Weapons and Foreign Policy* ([1957] 2018). However, the Nobel Peace Prize awarded in 2017 to the NGO-led International Campaign to Abolish

Nuclear Weapons suggests progress even in this bastion of hard power. The agreements mentioned in the preceding paragraph and this latest development in the nuclear arena are all signs of movement in favor of noopolitik's universalistic direction.

The relatively peaceful ending of the Cold War can be seen as another sign of a turn toward noopolitik. It was partly a case of the Soviet Union "going broke" because it was in a realpolitik-type arms race that it could not sustain. Yet, although Russia remained a major military power and political force in the world, then–U.S. President Ronald Reagan noticed that General Secretary of the Communist Party of the Soviet Union Leonid Brezhnev, though physically declining, seemed willing to engage in a competition of ideas and systems—particularly his view, stated in his Brezhnev Doctrine, that once a country turned socialist, it would never revert to capitalism.

So, in June 1982, in his "Address to Members of the British Parliament" (now known as the Westminster Address), Reagan outlined a hopeful vision in response, indeed a challenge, that smacks far more of noopolitik than realpolitik:

> The objective I propose is quite simple to state: to foster the infrastructure of democracy—the system of a free press, unions, political parties, universities—which allows a people to choose their own way, to develop their own culture, to reconcile their own differences through peaceful means. . . .
>
> Our military strength is a prerequisite to peace, but let it be clear we maintain this strength in the hope it will never be used, for the ultimate determinant in the struggle that's now going on in the world will not be bombs and rockets, but a test of wills and ideas, a trial of spiritual resolve, the values we hold, the beliefs we cherish, the ideals to which we are dedicated. (Reagan, 1982)

This speech became a foundational document for the revival of U.S. efforts to promote democracy abroad—first championed by former U.S. President Woodrow Wilson—leading in 1982 to creation of the National Endowment for Democracy and its associated party, labor, and business institutes, not to mention additional organizations in later years.[2]

After Brezhnev died in November 1982, his aging, unhealthy successors—Yuri Andropov and Konstantin Chernenko—strove to adhere to the Brezhnev Doctrine, but both died after short terms in office. The real difference in statecraft emerged in 1985 when Gorbachev, Chernenko's successor, and Reagan were able to join to forge a new, more noopolitik-like path. Indeed, Gorbachev's 1986 New Year's Day "Address to

---

[2]   For additional analysis of Reagan's Westminster Address, see the special series of articles by Richard Fontaine, Carl Gershman, and Daniel Twining, as well as a roundtable discussion hosted by Susan Glasser and David Kramer, in the November 2019 issue of *The American Interest* magazine.

the American People" is replete with notions of shared values and norms that radiated a clear sense of humanity's common purpose. In Gorbachev's words,

> [O]ur common quest for peace has its roots in the past and that means we have an historic record of cooperation which can today inspire our joint efforts for the sake of the future . . . Our duty to all humankind is to offer it a safe prospect of peace . . . We can hardly succeed in attaining that goal unless we begin saving up, bit by bit, the most precious capital there is—trust among nations and peoples. (Gorbachev, 1986, pp. 5–6)

The hopeful future path envisioned by Gorbachev appeared to be opening up at that time. By 1989, communism was crumbling in the Soviet satellite countries, brought down by people's civil-society movements, not violence—another early exercise in noopolitik. At the end of 1991, the Soviet Union itself dissolved. That same year, more than thirty countries had united to expel Iraq from Kuwait—signaling that overt military aggression was unacceptable—giving then–U.S. President George H. W. Bush fresh energy to repeat, in a speech to a joint session of Congress, a declaration he had first made in 1990 about a "a new world order" based on notions of personal, political, and economic freedom (Bush, 1991). The nascent information-technology revolution seemed to be enabling progress in all of these areas.

Then came the 9/11 attacks on America, revealing that not only civil society but also its dark flip side—"uncivil society," with its transnational terrorist, insurgent, and criminal networks—was gaining fresh empowerment from the information revolution, economic globalization, and the rise of new forms of organization. The ensuing War on Terror—often waged against other nations rather than specifically targeted at these dark networks—morphed into terror's war on the world. Terrorism has increased sixfold since 2001, and the number of ongoing wars, after decades of decline, has risen by one-third since 2010, from 30 to more than 40 at this writing (National Consortium for the Study of Terrorism and Responses to Terrorism, undated; Uppsala Conflict Data Program, undated). Meanwhile, Russia has fallen back on its more traditional authoritarian path, China has risen as a prospective great-power competitor, and roguish regional powers, such as Iran and North Korea, have focused on improving their hard-power capabilities. Realpolitik has come back into style.

Displacement of Realpolitik as the Noosphere Grows

Proponents of realpolitik might well prefer to stick with treating information as an adjunct to the standard political, military, and economic elements of diplomacy and grand strategy; the very idea of intangible information as a basis for a distinct dimension of statecraft seems antithetical to realpolitik. Realpolitik allows for information strategy as a tool of propaganda, deception, and manipulation but seems averse to treating "knowledge projection" as a true tool of statecraft.

For noopolitik to take hold, information must become a distinct dimension of grand strategy. The further development of soft power is essential for the emergence of this second path and thus of noopolitik. Without the emergence—and deliberate construction—of a massive, well-recognized noosphere, there will be little hope of sustaining the notion that the world is moving to a new system in which power is understood mainly in terms of knowledge (a point pioneered by Foucault). Thus, diplomats and other actors should start focusing on the balance of knowledge, as distinct from the balance of power.

Realpolitik, no matter how modified, cannot be transformed into noopolitik—the two stand in stark contradiction. This is largely because of the uncompromisingly state-centric, hard-power nature of realpolitik (see Bew, 2014). It is also because, for an actor to shift the emphasis of its statecraft from realpolitik to noopolitik, there must be a shift from power-maximizing politics to power-sharing politics. Nonetheless, the contradiction is not absolute; it can, in theory and practice, be made a compatible contradiction (rather like yin and yang). Indeed, true realpolitik often depends on the players sharing and responding to some core behavioral values—a bit of noopolitik might thus lie at the heart of realpolitik. Likewise, true noopolitik may work best if it accords, at least in some ways, with power politics. However, this perspective should be less about "might makes right" than about "right makes might." Understanding this might help in persevering through the transitional period in which realpolitik and noopolitik are likely to coexist. The point we reiterate for noopolitik is that this kind of world requires governments to learn to work conjointly with civil-society NGOs that are engaged in building transnational networks and coalitions.

Even now, many shifts, risks, and conflicts that are commonly categorized as geopolitical in nature are, on closer examination, primarily noopolitical. For example, during the past decade, the Arab Spring in the Middle East, the rise of the far right in Europe, Hindu-Muslim clashes in South Asia, and protest movements in Hong Kong are often written up as having geopolitical implications, yet they might be better understood as essentially noopolitical in nature. Indeed, myriad cognitive wars—ideological, political, religious, and cultural wars—are underway around the world, aimed at shaping people's minds and asserting control over this or that part of the emerging noosphere. At the same time, people are searching for new ways to get along and cooperate in addressing such global challenges as climate change and refugee settlement. Here, too, policies and strategies guided by noopolitik rather than realpolitik are bound to fare better for the common good of all parties.

Amid these challenges to the spread of the noosphere and the adoption of noopolitik, the United States—instead of leading the way for liberal societies to shift away from realpolitik—appears to be faltering badly, as we discuss next.

CHAPTER SIX

A Pessimistic Appraisal of Today's Turmoil for the Noosphere and Noopolitik

Our earlier work warned that, though some state and nonstate actors might find noopolitik attractive, they might care less about the emergence and construction of the noosphere. In the hands of a democratic leader, noopolitik might then amount to little more than airy, idealistic rhetoric with little or no structural basis, whereas, in the hands of a dictator or demagogue, noopolitik could be employed for purposes of manipulative propaganda and perception management. Narrower versions of noopolitik might also be fashioned for private gain; in the commercial worlds of branding, advertising, and public relations, this already occurs when companies concoct media blitzes and plant testimonials to spin public opinion. These were among the risks that might have to be faced, we warned long ago—and so did Colin Gray, in the only other forewarning we have found:

> The basis of the high regard Americans are inclined to have for soft power—aside from its low cost as compared with military force—lies in cultural hubris. It seems rarely to occur to us that we ourselves might be more vulnerable to civilizational co-option than are some others. (Gray, 2011, p. 53)

Unfortunately, our warning has been borne out, for noopolitik has been largely co-opted by dark actors. Today, despite its promise, noopolitik is not alive and well in the environment in which it should be thriving the most: the United States, where, now, even soft power is ailing as a strategic concept. Instead, state and nonstate adversaries—notably Russia and China, and, until lately, al Qaeda and the Islamic State (IS)—have developed their own versions of noopolitik, albeit by other names, and they have applied it effectively against the United States and its allies and friends. As noted earlier, these new circumstances mean that we are now living not only in the worst of times for noopolitik but also in the most pertinent—and urgent—of times for revisiting the promise of the noosphere and the prospects for noopolitik.

Washington Failing at Noopolitik

Not long ago, the United States (and individual Americans) had the lead in shaping the global noosphere, along with the ends, ways, and means it implies, at home and abroad. This is not true anymore, and has not been true since Washington began downplaying public diplomacy decades ago, since recent U.S. military engagements in the Middle East and South Asia caused ideational wear-and-tear, and since polarizing populism, political tribalism, and authoritarian-leaning nationalism have spread across the United States. America's "beacon" values and ideas—the American dream, the American experiment, American exceptionalism, America as a shining model of freedom and democracy, America as protector of the global commons—have become increasingly difficult, even controversial and questionable, to articulate and promote.

A few decades ago, America seemed to represent an "empire of ideas" (Hart, 2013). But now, Washington continues to pull back from wars of ideas, instead renewing an emphasis on the idea of war as the primary tool of foreign engagement—so much so that, as Andrew J. Bacevich wrote, "To cast doubts on the principles of global presence, power projection, and interventionism . . . is to mark oneself as an oddball or eccentric" (Bacevich, 2010, p. 27). Meanwhile, at home, American society has become so tribalized that ideas are being used more often as divisive weapons than as unifying agents. As many observers have noted, America no longer seems to have a unifying national story; indeed, it sorely needs to come up with a forward-looking new national narrative (e.g., Lepore, 2019). In the words of one astute observer,

> The question now is whether, as American power wanes, the international system it sponsored—the rules, norms, and values—will survive. Or will America also watch the decline of its empire of ideas? (Zakaria, 2019, p. 16)

For two decades, our prognosis has been that traditional realpolitik, which ultimately relies on hard (principally military) power, would increasingly give way to noopolitik, which relies on soft (principally ideational) power. But today's atmosphere is barely conducive to nurturing the noosphere or noopolitik. Washington seems bent on reaffirming guardedness over openness and hard power over soft power.

Indeed, Washington is floundering on soft-power matters. Today's mishmash of arguments about hard versus soft versus smart versus sharp types of power has muddled rather than clarified the bases for future U.S. strategy. Soft power remains a compelling concept, especially as a contrast to hard power. But the definition of *soft power* has been biased since the term's coinage decades ago (Nye, 1990; Nye, 2004). As we have discussed before (Arquilla and Ronfeldt, 2001, p. 349; Ronfeldt and Arquilla, 2007), its potential dark sides have received too little attention. The original definition tends to treat soft power as good and hard power as bad, or at least mean-spirited, for soft power is said to be fundamentally about persuasive attraction and hard power about coercion. But in actuality, soft power is not just about beckoning in attractive, upbeat, moralistic ways. It can be wielded in tough, dark, heavy ways, too, as in efforts to warn, embarrass, denounce, disinform, deceive, shun, or repel a targeted actor. Moreover, soft power does not inherently favor the good guys; malevolent leaders—e.g., an Adolf Hitler, an Osama bin Laden, or one of today's various authoritarian leaders—often prove eager and adept at wielding soft power in their efforts to dominate at home and abroad.

We warned back in 2007 that dark uses of soft power were "perhaps the most serious problem confronting American public diplomacy today" (Ronfeldt and Arquilla, 2007). It is reassuring, therefore, that a 2017 study from the National Endowment for Democracy (Walker and Ludwig, 2017a) shows that Moscow and Beijing have adroitly deployed *sharp power*—a dark variant of soft power—against the United States for some years now. Unfortunately, the authors' findings about Moscow's and Beijing's efforts were initially "dismissed" and met with "complacency" as the "the United States and other leading democracies seem to have withdrawn from competition in the ideas sphere" (Walker and Ludwig, 2017b, p. 9).

The authors urge, among other recommendations for fighting back, a reconceptualization of soft power because "The conceptual vocabulary that has been used since the Cold War's end no longer seems adequate to describe what is afoot" (Walker and Ludwig, 2017b, p. 23). We agree. Indeed, a rejoinder to their sharp-power concept errs in claiming that "Sharp power, the deceptive use of information for hostile purposes, is a type of hard power" (Nye, 2018). No, it is a dark variant of soft power. Perhaps the United States would not have proven so vulnerable to dark uses of soft power—thus, dark forms of noopolitik—these past few years if soft power (not to mention its spin-off, smart power) had not been viewed so benignly for more than two decades, thereby dulling U.S. wariness of new modes of cognitive subterfuge and aggression. It might well be correct that "Trump's presidency has eroded America's soft power," but it is not correct that soft power represents only "the power to attract rather than command" (Nye, 2020).

Making matters more ill-defined and uncertain, current American statecraft seems bent on exalting what may be termed *deal power*[1]—a narrow, transactional approach unsuited to long-range, worldwide, multilateral noosphere-building. Moreover, Washington seems to be backing away from having a grand strategy and strategic narrative that are forward looking about where the world should be heading—which means that the United States is ceding noospheric ground to visions that U.S. state and nonstate adversaries are promoting. Conservative arguments have even appeared questioning whether soft power is particularly effective nowadays (Dinerman, 2019). All of this is most perilous and does not augur well for noopolitik's broader adoption.

---

[1] The term *deal power* is our coinage for representing U.S. President Donald J. Trump's penchant for making deals, as he expressed in his 1987 book with Tony Schwartz, *The Art of the Deal*.

Besides pulling back from promoting traditional American values abroad, Washington has allowed setbacks in its commitment to science. Remember, the noosphere began as a scientific concept—Teilhard, Le Roy, and Vernadsky all were serious scientists concerned about the future of Earth. But in today's America, we see mounting attacks on science and scientific research that enable new commercial and governmental exploitations of the geosphere, the biosphere, and ultimately the prospects for the noosphere. In recent decades, Washington-based setbacks have ranged from denying the findings of tobacco research to outlawing gun research, discrediting climate research, and restricting what kinds of public health and environmental studies can be used by policymakers. In many controversial issue areas, science has become polarized and politicized (e.g., see Friedman, 2019; Meyer, 2020; Plumer and Davenport, 2019).

Perhaps the iconic examples of Washington's current disregard for noospheric initiatives are its withdrawals from the Paris Agreement on climate change and the Trans-Pacific Partnership on trade and investment in 2017. These moves represented a turning away not only from liberal internationalism but also from the promise of noopolitik. Nonetheless, various American actors—notably in military and civil-society circles concerned about the global commons—are still quietly working on pro-commons noospheric initiatives that might reenergize the rise of noopolitik. We discuss that later. But first, Washington and its friends and allies need to face a more urgent challenge: the resort to dark forms of noopolitik by state and nonstate adversaries.

Moscow, Beijing, Tehran, and WikiLeaks: Noopolitik's Darker Champions

The soft-power concept undergirds the noopolitik concept. Unfortunately, as discussed, the proponents of soft power have long defined it as something bright and attractive, making hard power look dark and coercive; they have ignored the dark uses to which soft power can be put. However, as we have recognized from the beginning, noopolitik can be put to dark purposes. Indeed, each of America's challengers for global influence has shown considerable aptitude for noopolitik. Moscow has resorted to longstanding political-warfare techniques, finding them highly advantageous for subverting democracy. Beijing is working on expanding its global reach via a mix of soft- and hard-power initiatives. Tehran is fashioning its expansionist strategy around its Shi'ite coreligionists, basing its kind of noopolitik on a religious noosphere. In addition, various nonstate actors, notably WikiLeaks, have employed dark forms of noopolitik to attack liberal values and interests.

What are these darker forms? They go by many names: *information warfare, information operations, cognitive warfare, political warfare, memetic warfare, epistemic warfare, neocortical warfare, perception management,* and *strategic deception,* along with such older terms as *the war of ideas* and *the battle for hearts and minds* and newer terms about weaponized social networks and weaponized narratives. What these terms have in common is that they all represent ways to work on the mind—sometimes for good, other times for ill. By way of contrast, we view noopolitik as a way to work with the mind.

For a while, nonstate adversaries—notably, al Qaeda and IS—seemed to lead in mastering the arts and techniques of cognitive warfare. But they no longer pose the dire, imminent threats of a few years ago. Now, it is various nation-state adversaries that have the lead in using dark varieties of noopolitik.

In the case of Russia, this means influence operations that go by such names as *Active Measures, kompromat, dezinformatsiya, reflexive control*, and *hybrid warfare*. Moscow uses them to deploy strategic narratives that extol "Eurasianism" and "traditionalism" and disparage the decadent West and democracy in ways that divide Americans and Europeans from each other. A penchant for online raiding and brigandage appears to underlie Moscow's approach to political warfare against the West, for "Cyber operations are perhaps the most obvious instrument for modern day raiding" (Kofman, 2018). To such ends, Moscow has invested heavily in radio and television stations, social media platforms, and internet troll farms (McFaul, 2018). One online platform, the Internet Research Agency, "was able to masquerade successfully as a collection of American media entities, managing fake personas and developing communities of hundreds of thousands" in efforts "to manipulate and exploit existing political and societal divisions" (DiResta, 2018b; see also Jenkins, 2018). No culture-war issue area is too fringe for Moscow—its efforts have even extended into right-wing home-schooling networks in the United States (Michel, 2019).[2]

The psychological theory behind Moscow's version of political warfare is deeply rooted in Russian history. Grigory Potemkin used deception and disinformation in the late 18th century to make people (notably, Empress Catherine II) think things were better, or different, than they actually were, by creating so-called Potemkin villages—as seen today in online "Potemkin villages of evidence" that support Russian influence operations (Pamment et al., 2018). Another foundational reference point is Ivan Pavlov's psychological work on reflexive conditioning in the late 19th century, resulting in Pavlovian conditioning.[3] To say that Russian strategy has "Pavlov-ed and Potemkin-ed" many American minds might sound odd, but it might well be accurate.

Yet a third reference point from Russian history stems from information operations developed in the 1920s on Joseph Stalin's behalf in Moscow's State Political Directorate (GPU, the precursor to the KGB) to develop deception and disinformation as deliberate tactics, culminating in efforts to create arresting atmospheres abroad, as

---

[2]   Other sources on Russia include Schoen and Lamb, 2012; Pomerantsev and Weiss, 2014; Pomerantsev, 2015; Lucas and Nimmo, 2015; Walker and Ludwig, 2017a; Lynch, 2018; Allen and Moore, 2018; Hwang, 2019; Pomerantsev, 2019; Graham, 2019; and Lucas, 2020.

[3]   Somewhat related is *operant conditioning*, a learning concept developed by B. F. Skinner, beginning in 1938.

well as at home, in which nothing seemed reliably true—a de-truthing so profound that it corresponds to what Peter Pomerantsev, 2019, calls a "propaganda of unreality" built around a "futureless now."[4] Adam Ellick, a specialist on Russian disinformation operations, notes that

> Today Russia is launching a disinformation campaign against the States, which is basically trying to get us to believe in nothing. It's trying to erode belief and credibility in anything. And it's also us against us. And this country is so split and divided that we're now using this Soviet playbook, this disinformation playbook, on ourselves (Gross, 2018).

Decades ago, from the 1920s through the 1980s, when communism and the communist party still possessed significant sway around the world, Moscow focused almost entirely on subverting and co-opting actors on the Left, particularly in the United States and Europe. Even as Gorbachev sought accommodation with the West, Soviet "active measures" persisted in the pursuit of Soviet goals among actors on the Left. As Herbert Romerstein, at the time a senior director in the United States Information Agency, put it, the Soviet Union under Gorbachev had "not changed its view of the world but only its tactics" (Romerstein, 1990, p. 38).

But since the West seemingly won the Cold War, and especially since Putin's rise, Russian and affiliated information operations have focused more on subverting and exploiting actors on the Right. They have done this by energizing such issues as nationalism, traditionalism, racism, and sexism, not to mention all sorts of conspiracy theories, to make Western societies more fractious than ever (see Applebaum, 2019; Herf, 2019).

As for China, Beijing arguably is not as openly antidemocracy and pro-autocracy abroad as Moscow, being more interested in regional primacy than global hegemony. But Beijing is fielding strategic narratives that call for a new type of great-power relations, a new international system, and a new Chinese-like development model ("socialism with Chinese characteristics"), all to create a "world organized as a network" (Deng Xiaoping's vision) and a "community of common destiny with mankind" (Xi Jinping's vision)—in sum, a "Chinese Dream" to rival the American dream (Li and Xu, 2014; Mardell, 2017). Beijing's enticing (but extractive) grand strategy includes its Belt and Road Initiative, the deployment of Confucius Institutes around the world, the creation of political party cells (Chinese Students and Scholars Associations) wherever Chinese students are studying abroad, the spread of the Chinese People's Association for Friend-

---

[4] See also Morson, 2019, on "Leninthink."

ship with Foreign Countries, and investments in U.S. and other foreign think tanks—all serving to extend a networked Chinese noosphere through influence operations.[5]

These and other stealthy soft-power (or sharp-power) projections reflect a Chinese approach to cognitive influence known as *Three Warfares*—a way to use public-opinion, legal, and psychological operations in combination (Mattis, 2018). One major element is *huayuquan*—a cognitive warfare concept about "the capability to control the narrative in a given scenario," or "discourse power" (Kania, 2016; Kania, 2018; see also Doshi, 2020; Gregory, 2018a; Gregory, 2018b; Hwang, 2019).

Indeed, asserting discourse power at home and abroad has become a crucial element in Beijing's strategy, as shown in this string of observations from leading authorities:

> This focus on discourse power highlights that the CCP [Chinese Communist Party] and PLA [People's Liberation Army] recognize the power of narrative as a means to achieve advantage relative to a rival or adversary through reshaping the information environment in ways that have real-world impact. (Kania, 2018)

> China's global media acquisitions and operations, efforts to dominate the Chinese diaspora, influencing and manipulating the politics of liberal-democratic nations and expanding influence in international organizations are all indicative of China's international 'discourse power' project. (Gregory, 2018b)

> China refers to its attempt to control the narrative about China as a war, a *huayuzhan* . . . or a discourse war. (Diamond and Schell, 2019, p. 100)

> Many of the threats to the CCP and its political system occur in the realm of ideas. They cannot be defeated by kinetic means. . . . [T]hreats must be preempted in the minds of foreign policymakers who might choose to compete, contain, or attack China. (Mattis, 2018)

> Xi is exporting his brand of colonialism with Chinese characteristics through soft power means, on a road toward achieving his hard power ends. (McCaul, 2018)

> And if Beijing can leverage its growing control over the information supply chain to quell reporting on its corruption, elite capture, and information campaigns in other countries, it will damage U.S. interests and pose a significant risk to democratic accountability around the world. (Doshi, 2020)

---

[5]   Other sources on China include Barno and Bensahel, 2016; Campbell and Sullivan, 2019; Diamond and Carson, 2019; Edel and Brands, 2019; Patrikarakos, 2019; Qiao and Wang, 1999; Swaine, 2019; Weiss, 2019; and Zakaria, 2019.

Overall, then, the U.S.-China relationship has become so fraught with "huge potential risks and dangers" that it now "needs to develop a new strategic narrative" (Swaine, 2019).

What matters for our work is that Beijing appears to be taking noosphere and noopolitik seriously, though by other names. Indeed, China's concept of discourse power resembles our noopolitik more than any other concept we have come across. China is taking its concept so seriously, and treating it so systematically, that Washington risks being outmaneuvered partly because of that discrepancy in focus.

Meanwhile, myriad nonstate actors are using dark forms of noopolitik far and wide, not only on the extreme Right and Left politically but also in areas and on topics that are not easily categorized. Prominent examples have included al Qaeda, IS, QAnon, and WikiLeaks, not to mention various other nationalist, religious, and ideological actors around the world who want to target the United States. Most have their own agendas and operate independently, but some might also act on occasion as proxies for state or corporate actors.

These nonstate actors have become enormously adept at waging cognitive warfare via darkly mutant kinds of noopolitik, both online and in real life. They have learned to create and deploy strategic narratives constructed around "plausible promises," "sacred values," and "viral conspiracy theories" that serve not only to attract recruits, enliven morale, and stimulate network building within and among their own kind but also to confront, disorient, divide, and destabilize their targets.[6] More and more, their media campaigns exhibit the netwar-swarming methods that we predicted two decades ago (see Arquilla and Ronfeldt, 2000; Arquilla and Ronfeldt, 2001).

The information-age ways that these actors conduct cognitive warfare do not depend on coercive brainwashing, mind control, or thought reform—concepts that came into play during World War II and the Korean War, partly as a result of psychological techniques born of the Russian and Chinese revolutions.[7] Yet today's dark ways do indeed aim to achieve what amounts to *thought-scripting* and *brain-boxing*. With the former, a person's thinking comes to congeal around increasingly set scripts that the person's mind runs (and voices) over and over. This goes beyond a common tendency to tell the same story again and again; it makes minds more reactive and programmed, as though a cognitive button is pushed or a lever pulled, and out pops a set script, positive or negative. And if the script is arguable, it is unlikely to be changed through argument.

With brain-boxing (or thought-boxing, or mind-boxing), one's thinking about the world becomes increasingly boxed within a frame (see Lakoff, 2014, on framing).

---

[6]   On plausible promises, see Robb, 2007; Shirky, 2008. On sacred values, see Atran, Sheikh, and Gómez, 2014; Francis, 2016. On viral conspiracy theories, see Butter and Knight, 2020; Goldenberg and Finkelstein, 2020.

[7]   The Russians, in particular, developed these concepts extensively, for the most part under the rubric of their theory of "reflexive control." See Thomas, 2004.

What people think, and how they think, about the world—their world—becomes increasingly fixed, enclosed, and bounded. Scripts then run within that frame and its boundaries. The well-boxed brain rarely goes looking for new ideas and topics to think about; it prefers reassurance and reinforcement about what is already in the box. Brain-boxing is another way by which people become set in their ways. These days, the minds of extremists might not exhibit brainwashing, but brain-boxing is another story—and it is a story that reflects not only current Russian and Chinese practices but also much of American advertising, news, and entertainment programming.

The more tribalized a mind, and the more polarized the political environment, the more public dialogue will exhibit such scripting and boxing. This will make efforts to promote balance and compromise not only unattractive but well-nigh impossible. What is at stake, then, is not just whose story wins but also, and more ominously, whose emergent noosphere gains ground.

The Noosphere in Fragmented Disarray

What would a full-fledged noosphere encompass? What ideas, values, and norms— what principles, practices, and rules—should it embody? We presume that these would include much that the United States and its friends and allies stand for: openness, freedom, democracy, the rule of law, humane behavior, respect for human rights, a preference for peaceful conflict resolution, etc.—all that the noosphere's original pro-ponents said should and would be embedded. In addition, a full-fledged noosphere would require an interactive organizational and technological foundation to uphold its ideational essence.

However, the world is not yet in the age of the noosphere, but rather in an era of transition that is far from smooth or peaceful (as Vernadsky, in particular, predicted). When we first started writing about the noosphere and noopolitik, we figured we were witnessing halting steps forward. Yet the steps backward are what are most evident today, especially in the behavior of some of the world's most powerful states, as we have just discussed.

No methodology exists for assessing the status of the noosphere from strategic standpoints; nobody has yet seen to that potentially valuable task. Even so, we can observe that the noosphere is in terribly fragmented disarray in the very country, the United States, that should be taking the most initiative to uphold and foster it.

Much of the evolving liberal noosphere has become highly compartmentalized— broken up into what are often called *information silos*, *filter bubbles*, and *echo chambers*, tantamount to volatile microclimates. Many of these "compartments" and "cultural units" (Teilhard's terms) are engaged in "ruthless struggle" (Vernadsky's words), far from being ready for the "fusion" that Teilhard forecast or to "open up and finally link up their spouts, spreading a layer that covers the Earth," as Le Roy depicted (Samson and Pitt, 1999, p. 66). Indeed, the noosphere is presently so fragmented, and many of its "units" so polarized and tribalized, that it could be said that a war—a cultural civil war rife with "memetic tribalism" (Limberg and Barnes, 2018)—is underway for control of the noosphere. And this war reflects the essence of noopolitik, for it will be decided by whose story wins.

For example, the gun rights advocacy of the National Rifle Association (NRA) is not just about guns anymore, but rather about defending a way of life, a system of beliefs and principles, a kind of tribal culture. Its leaders, members, and supporters are tightly networked, in organizational and media terms. If criticized or otherwise attacked rhetorically, they rarely waiver—they have value orientations and memetic reflexes at the ready. Altogether, the NRA is like a tribal mini-noosphere with a hard shell, though it might be starting to crack because of internal leadership conflicts (Melzer, 2009). More to the point, similar platforms, in one issue area after another, are spread all across America, some with transnationally networked ties to compartmentalized entities elsewhere.

Admonitions from the Edges

By now, we are far from alone in trying to warn about the disarray threatening the current phase in the noosphere's evolution. Here, albeit without explicit reference to the noosphere, is a dramatic parallel warning from Renée DiResta:

> There is a war happening. We are immersed in an evolving, ongoing conflict: an Information World War in which state actors, terrorists, and ideological extremists leverage the social infrastructure underpinning everyday life to sow discord and erode shared reality. The conflict is still being processed as a series of individual skirmishes—a collection of disparate, localized, truth-in-narrative problems—but these battles are connected. The campaigns are often perceived as organic online chaos driven by emergent, bottom-up amateur actions when a substantial amount is, in fact, helped along or instigated by systematic, top-down institutional and state actions. This is a kind of *warm war*; not the active, declared, open conflict of a hot war, but beyond the shadowboxing of a cold one. (DiResta, 2018a, emphasis in original)

As a result, Americans in particular are deep into an "an ongoing battle for the integrity of our information infrastructure," such that "the norms that have traditionally protected democratic societies will fail" (DiResta, 2018a). How it ends could be "as consequential in reshaping the future of the United States and the world as World War II" (DiResta, 2018a).

Likewise, Jonathan Rauch, 2018, observes that our political culture is under an "*epistemic* attack . . . on our collective ability to distinguish truth from falsehood"—

and it is affecting "the constitution of knowledge" that enables a society to function in an organized manner.[8] In another recent warning about the implications for democracy, Larry Diamond observes that

> The world is now immersed in a fierce global contest of ideas, information, and norms. In the digital age, that contest is moving at lightning speed on an hourly basis, and it is shaping how people think about their political systems and the future world order. Now especially—when doubts and threats to democracy are mounting in the West—this is not a contest that the democracies can afford to lose. (Diamond, 2019)

Matters are so fraught that, according to a recent RAND report, "The United States . . . needs an updated framework for organizing its thinking about the manipulation of infospheres by foreign powers determined to gain competitive advantage" (Mazarr, Casey, et al., 2019, p. xii). We fully agree. We further agree with their follow-up point that the United States remains vulnerable to emerging modes of "virtual societal warfare" waged largely in and through the infosphere, mostly by way of networks fighting networks (Mazarr, Bauer, et al., 2019). Improving our defenses will require making that infosphere more resilient. But unfortunately, as a third RAND report shows, political and civil discourse in the United States is increasingly characterized by "Truth Decay" that makes problem-solving all the more difficult (Kavanagh and Rich, 2018).

None of this bodes well for the noosphere or for the noopolitik we have proposed—it is all grounds for pessimism. But although the foregoing turmoil dominates the news and other media, shaping people's impressions and reflexes about what is going on around the world, we have found a new reason to sustain our hopes for the future of the noosphere and noopolitik—a hope we share in the next chapter.

---

[8]   On "network propaganda" inducing "epistemic crisis," see Benkler, Faris, and Roberts, 2018. See also Singer and Brooking, 2018.

CHAPTER SEVEN

Hope for the Noosphere and Noopolitik: The Global Commons

Our writings have stressed that noopolitik, far more than realpolitik, may depend on close cooperation among state and nonstate actors. In particular, we have pointed out the important roles that networked civil-society NGOs can play. Thus, we noted early cases of NGOs successfully using noopolitik—e.g., the International Campaign to Ban Landmines, a coalition of NGOs that won the Nobel Peace Prize in 1997. And we have listed a variety of issue areas where state-nonstate cooperation can help foster the noosphere and noopolitik: e.g., human rights, conflict resolution, democracy promotion, and the environment.

To this list, we now add the *global commons*—traditionally, the parts of Earth that fall outside national jurisdictions and to which all nations are supposed to have access, such as the high seas, the atmosphere, and outer space. The global commons might turn out to be a pivotal issue area.

Although the noosphere and noopolitik are not faring well in the power centers discussed in the prior chapter, the noosphere concept is progressing better among actors around the world who are concerned about the global commons. This concept is thus of keen interest because it relates closely to the prospects for the noosphere. Moreover, actors concerned about the security and accessibility of the global commons seem naturally attracted to noopolitik.

Indeed, it may well turn out that the noosphere and noopolitik concepts will fare better in the future the more they become associated with the concept of the global commons—and the latter might flourish the more it can be associated with the noosphere and noopolitik. Recognizing the noosphere's association with the global commons might then help put noopolitik back on track in various strategic issue areas, despite the aforementioned negative trends.

What makes the global commons concept potentially pivotal is that it has taken hold in two seemingly contrary circles. One is civilian, forming mainly at the behest of IGOs, NGOs, and other nonstate actors motivated by environmental and social matters. The other circle is military, motivated by state-centric security interests. Furthermore, although the term *commons* has been used for centuries, *global commons* is quite recent. It first appeared in civilian environmental circles—implicitly in nego-tiations behind the United Nations Convention on the Law of the Sea during 1973–1982, and then explicitly in the Brundtland Commission's 1987 report, *Our Common Future* (World Commission on Environment and Development, 1987). In the fol-lowing decades, the term spread into military and security strategy circles, notably in the 2008 U.S. *National Defense Strategy* (U.S. Department of Defense [DoD], 2008), and then to greater effect in the 2010 U.S. *Quadrennial Defense Review Report* (DoD, 2010). Both the civilian and the military views about the global commons became espe-cially important during the administration of former U.S. President Barack Obama (Ikeshima, 2018; Li, 2012).

The global commons is thus bracketed by differences in its meanings in environmental-science and civil-society circles on the one hand and in military circles on the other. In the past, these different circles rarely interacted; some pro-commons civil-society activists even objected to seeing the term show up in military circles (Bollier, 2010; Morris, 2011). Now, however, as more actors in both civilian and military circles recognize the adverse effects of climate change and other global environmental shifts affecting the carrying capacity of the planet, the views held in these seemingly contrary circles might start to intersect, as may their separate calls for reforms and remedies. As former Chairman of the U.S. Joint Chiefs of Staff Admiral Michael Mullen stated about climate change, its "potential impacts are sobering and far-reaching" (Berlin, 2013). Evidence of such emerging awareness in national security circles suggests the prospects for the noosphere and noopolitik may vastly improve.

In this chapter, we first discuss perspectives in environmental-science and civil-society circles. Next, we discuss military perspectives on the global commons. Finally, we highlight their intersections and implications for policy and strategy, particularly for nurturing noopolitik.

Environmental-Science and Civil-Society Views on the Global Commons

Civilian interest in the global commons involves two different circles. One consists of scientists and associated actors (international organizations in particular) who are concerned about environmental matters. They have grown into a large, influential community (or set of circles) and have billions of dollars at their disposal. The other circle consists largely of pro-commons civil-society activists whose agendas include not only environmental issues but also the radical transformation of societies as a whole. This community is growing around the world too, though in a low-key, low-budget, bottom-up manner.

The two have much in common, but they are also distinct: The big-environmental-science circle generally seeks to have government, banking, business, civil-society, and other actors work together to protect the biosphere. It mostly leans in progressive and liberal internationalist directions and calls for societies to achieve transformational change in certain environmentally sensitive sectors, but mostly through "our new way of doing business." In contrast, the social-activist civil-society circle is decidedly of the Left—but it is part of a new kind of Left, for it wants "commons-based peer production" and other kinds of "commoning" to spread to such an extent that societies undergo a phase shift away from capitalism to new commons-based forms of social organization. The members of this circle have far more on their minds than environmental science and the biosphere.

The Big-Science Community
The biggest advances in thinking about the global commons come from scientists and related actors focused on global environmental matters. They take the biosphere concept very seriously (and at times allude to the noosphere or Gaia). And they have formed into a worldwide network of IGOs, NGOs, research centers, private individuals, and government, banking, and business actors—with the United Nations Environment Programme (UNEP), the Global Environment Facility (GEF), and the recently created Global Commons Alliance serving as key network hubs. Moreover, the GEF is fostering a growing Movement of the Global Commons (or, Movement for the Global Commons) that aims to "develop a compelling story about needs and opportunities for the Global Commons" and engage people "from communities to corporations to cabinets" (GEF, 2017, p. 10; see also GEF, undated; GEF, 2019a; GEF, 2019b; Global Commons Alliance, 2019; and UNEP, undated).

Several decades ago, environmental concerns were mainly about specific local matters, such as air or water pollution. Late in the 20th century, after decades of seeing problems worsened by "global forces of consumption, production, and population," environmentalists realized that their challenge was planetwide, involving what they began calling "the global commons"—meaning "the shared resources that no one owns but all life relies upon" (Levin and Bapna, 2011, p. 30). As this concept took hold, mostly after the Brundtland Commission's report in 1987, its proponents came to identify the high seas, the atmosphere, Antarctica, and outer space as the domains of interest—and they did so "guided by the principle of the common heritage of mankind" and a sense of "common responsibilities" (United Nations System Task Team on the Post-2015 UN Development Agenda, 2013, pp. 5–6). These concepts would make for considerable overlap with the later military view that the global commons consist of four operational domains: sea, air, space, and cyber (in contrast, the concepts about mankind's common heritage and common responsibilities were quickly criticized and resisted by multinational business corporations).

Some proponents have sought to expand the global commons further. Thus, "Resources of interest or value to the welfare of the community of nations—such as tropical rain forests and biodiversity—have lately been included among the traditional set of global commons . . . while some define the global commons even more broadly, including science, education, information and peace" (United Nations System Task Team on the Post-2015 UN Development Agenda, 2013, pp. 5–6). Proponents for including biodiversity often mention preserving the quality of soil and marine conditions. Such views would mean expanding the global commons in social directions that are most pronounced within the civil-society circles discussed in the next section of this chapter.

These various proponents (notably, Nakicenovic et al., 2016; Rockström et al., 2009b) urge viewing the global commons and "the large-scale subsystems of the Earth system—ocean circulations, permafrost, ice sheets, Arctic sea ice, the rainforests and atmospheric circulations"—as a tightly coupled, complex system characterized not only by stable equilibria but also by "regime shifts, tipping points, tipping elements, nonlinearities and thresholds" that may generate a "bifurcation point" and then lead to "a new equilibrium state" or sudden collapse (Nakicenovic et al., 2016, p. 17). The threat is that

> If one system collapses to a new state, it may set up positive feedback loops amplifying the change and triggering changes in other subsystems. This might be termed a 'cascading collapse' of key components of the Earth system. (p. 17)

As discussed later, this kind of analysis matches how the U.S. military has come to view the four domains comprising its global commons as a complex interactive system.

Of key conceptual importance for the big-science perspective, Johan Rockström, director of Sweden's Stockholm Resilience Center, has provided seminal studies for years about biosphere interactions and planetary life-support systems. He formulated crucial new concepts about "planetary boundaries" (Rockström et al., 2009a)—in particular, "nine planetary boundaries [that] provide a safe operating space for humanity" (Rockström, 2011). In his view, several boundaries have already been transgressed, and further slippage looms. Accordingly, humanity threatens to cause catastrophes that can overwhelm the biosphere and thus the Anthropocene. Indeed,

> The high seas, the atmosphere, the big ice sheets of the Arctic and Antarctica, and the stratosphere—traditionally seen as the Earth's global commons—are now under suffocating pressure. Yet we all depend on them for our wellbeing. (Rockström, 2017; see also Nakicenovic et al., 2016; Rockström et al., 2009b, Rockström, 2011)

As a result, not only is further scientific research needed but also new global perspectives, narratives, organizations, and strategies to assure planetary resilience, sustainability, and stewardship—if possible, to achieve a holistic transformation. According to Rockström, 2011, p. 21, "Governance of the global commons is required to achieve sustainable development and thus human wellbeing. We can no longer focus solely on national priorities." Looking farther out, Nakicenovic et al., 2016, p. 27, insists that "all nation states have a domestic interest in safeguarding the resilience and stable state of all Global Commons, as this forms a prerequisite for their own future development." Therefore,

> Stewardship of the Global Commons in the Anthropocene, with its three central principles of inclusivity, universality and resilience, is an essential prerequisite to guide national and local approaches in support of the Sustainable Development Goals for generations to come. (p. 46)

Rockström, 2017, goes so far as to predict that if the right steps could be taken on behalf of the global commons, then "planetary intelligence could emerge on Earth by 2050." His language sounds much like that of Teilhard and Vernadsky—but falls just short of explicitly mentioning the noosphere:

> Here's a prediction: planetary intelligence could emerge on Earth by 2050.
>
> . . . planetary intelligence emerges when a species develops the knowledge and power to control a planet's biosphere. . . .
>
> For planetary intelligence to emerge on Earth within three decades we need to change our worldview, our goals and our rules.
>
> . . . we must redefine the global commons. In these new circumstances we can now define them as a resilient and stable planet. That is every child's birthright, and our common heritage; but it is now at risk. The Anthropocene and the new global commons represent a new worldview—a paradigm shift—as fundamental as Darwin's theory of evolution or Copernicus's heliocentricity. (Rockström, 2017)

As for steps yet to be taken, Rockström and many of his colleagues believe that "[w]e desperately need an effective global system of governance" (Rockström, 2017, p. 25). The concern is that "In a period of increasing interdependence and complexity, global governance remains fragmented, hampered by loud national interests, and unable to address global risks that present non-linear dynamics and repercussions" (Rockström, 2016, p. 121). The following are needed: new legal norms about planetary boundaries; a stronger role for UNEP; stronger commitments by "governments, private actors and the international community" to adopt innovations to safeguard the biosphere; along with "a recognition that transformative change requires engagement and mobilization 'from below' . . . endorsed by the population" (Rockström, 2016, p. 121). And while the work of Rockström and his colleagues focuses mostly on defining thresholds and rights for using the global commons, other work by the Global Thresholds and Allocations Council aims to define fair allocation mechanisms in a "partnership between leading organizations and individuals from science, business, investment, government, and civil society" ("Global Thresholds and Allocations Council (GTAC)," 2017; see also Baue and Thurm, 2018).

Meanwhile, in light of the scientists' concept of nine planetary boundaries, economist Kate Raworth has added a set of eleven "social boundaries" below which income, education, food, water, and other such levels should not be allowed to fall. She has assembled all of these boundaries into "a visual framework—shaped like a doughnut" (Raworth, 2012, p. 1). The planetary boundaries are depicted in an outer circle, the social boundaries in an inner one. As a result,

> Between the two boundaries lies an area—shaped like a doughnut—which represents an environmentally safe and socially just space for humanity to thrive in. It is also the space in which inclusive and sustainable economic development takes place. (Raworth, 2012, p. 4; see also Raworth, 2017; Rockström and Raworth, 2015)

Raworth's writings about "doughnut economics" have added clarity to the implications of the planetary-boundaries concept—but, just as significant, the ways she relates her ideas to the global commons and to the prospects that future societies will contain a commons sector have spread her influence into the pro-commons civil-society social-activist circles discussed next.

Providing a boost to all of the aforementioned are efforts by the GEF under the leadership of Naoko Ishii. Before Ishii became chief executive officer and chairperson in 2016, GEF had been "producing numerous, small projects in a fragmented or isolated way without in total really shifting the needle in the right direction or triggering transformational change" (GEF, 2016). So she designed a new long-term strategy to focus on "the drivers of environmental degradation," as well as on its ecological and social effects. Impressed by the "planetary boundaries" framework, she too has come to insist that "Everything is connected . . . If one area is in deep trouble it will affect others and lead to disruption to the whole planetary system" (GEF, 2016). Furthermore, she observes,

> Now our economies have become global, but the principle of taking care of common resources has not. The whole world faces the tragedy of the commons. The global commons, on which all humanity depends, are being pushed towards breaking point, posing an ever-increasing threat to our aspirations for economic growth, jobs and security. . . .
>
> We need a new way of doing business. We must recreate the same kind of contracts that local communities found so effective in the past. But we must do so on a global scale—building new coalitions and partnerships to transform the key economic systems that support how we eat, how we move and how we produce and consume. (Ishii, 2019)

Again, these and other points discussed in this section resemble points made by military proponents of the global commons, as discussed later.

Social-Activist Networks

For the military, the sea was the first global commons. But for civil-society activists, the concept of the commons originated far earlier, first as an ancient Roman distinction between *res nullius* (something owned by no one) and *res communis* (something owned in common), and then later in medieval England as a way to refer to open land "held in common."

By now, according to pro-commons civil-society theorists and activists, the concept includes not only natural physical commons—land, air, and water, as "gifts of nature"—but also digital commons (online terrain and knowledge as commons). More than that, some activists include social commons—e.g., cooperatives, where creative work amounts to a shared asset. Culture is sometimes viewed as belonging to the commons as well.

Pro-commons proponents in civil-society circles define *commons* as shared resources co-governed by a community (users and stakeholders) according to the rules and norms of that community. All three components—resource, community, rules— in other words, the "what," the "who," and the "how"—are deemed essential. Together, they mean "the commons" is not just about resources or terrain; it is about a way of life called "commoning." Furthermore, an eventual aim of many "commoners" is to create a new "commons sector" (Bollier and Rowe, 2006) alongside, but distinct from, the established public and private sectors. If and as this develops, a revolutionary societal transformation will occur. Indeed, a goal of some pro-commons theorists and activists is to "build 'counter-hegemonic' power through continuous meshworking at all levels" so that "the destructive force of global capital and its predation of the planet and its people can be countered" (Bauwens et al., 2017, p. 42; see also Bauwens, Kostakis, and Pazaitis, 2019; Bauwens and Ramos, 2018; Ronfeldt, 2012).

Fifty years ago, the commons concept had little traction in advanced societies— especially after Garrett Hardin famously wrote "The Tragedy of the Commons" (1968). Today, however, pro-commons social movements are thriving around the world. They were inspired initially by people experiencing the internet and the World Wide Web as a kind of commons, even as a harbinger of the noosphere. Then, Elinor Ostrom's book *Governing the Commons: The Evolution of Institutions for Collective Action* (1990) and her Nobel Prize in economics in 2009 aroused many people to realize, contrary to Hardin and other critics, that common-pool resources can indeed be managed productively (Ostrom, 1990; Stavins, 2011; Wilson, 2016). By now, pro-commons movements are slowly, quietly expanding throughout North America, Western Europe, and Scandinavia, gaining inspiration and guidance from a host of new civil-society NGOs, notably the P2P Foundation led by Michel Bauwens, as well as from individual theorists, such as David Bollier and Yochai Benkler. In some instances, further impulse comes from

Green political parties and movements. In comparison with the big-environmental-science circle, this is not a hugely influential circle (yet), but it is generating a social movement that is raising interest in the global commons and the noosphere.

Much of this innovation is occurring on the Left. German commons advocate Silke Helfrich (quoted in Bollier, 2014) has noted accurately that "commons draw from the best of all political ideologies"—for example, from conservatives, the value of responsibility; from liberals, the values of social equality and justice; from libertarians, the value of individual initiative; and from leftists, the value of limiting the scope of capitalism. Yet this is still largely a set of movements from left-leaning parts of the political spectrum. So far, few conservatives have realized the potential benefits of allowing a commons sector to emerge. Indeed, on the Right, resistance to pro-commons ideas and advocates is a regular theme—from "America First," to Brexit, to the nativist Alternative for Germany political party, and others.

At first, two or three decades ago, pro-commons activists focused primarily on local and national matters. But as visions have evolved, more and more activists have redirected their focus beyond local and national commons and toward expansive global concepts. This turn is well underway. For example, German economist Gerhard Scherhorn, 2012, would include in the global commons not only natural resources but even "employment opportunities, public health systems, educational opportunities, social integration, income and wealth distribution, and communication systems such as the Internet." A further example is the analysis of international development expert and commons advocate James Quilligan that

> While watching markets and states mismanage the world's cross-boundary problems, it has dawned on many individuals, communities and civil society organizations that the specific objectives we are pursuing—whether they are food, water, clean air, environmental protection, energy, free flow of information, human rights, indigenous people's rights, or numerous other social concerns—are essentially *global commons* issues. (Quilligan, 2008, emphasis in original)

Meanwhile, many pro-commons civil-society proponents on the Left seek organizational changes that resemble those from the big-science and military circles. In this vein, Quilligan, 2008, proposed "that we would gain considerably more authority and responsibility in meeting these problems by joining together as *global commons organizations*" (emphasis in original). Accordingly, "The challenge is to assemble international representatives from all regions and sectors to discuss global commons issues in a negotiating format which integrates these three streams [i.e., geosphere, biosphere, noosphere] of evolution" (Quilligan, 2010, p. 47). Quilligan, like others, has recommended that local communities of users and producers agree to new kinds of "social charters" and "commons trusts" to assure their hold on commons property. If more and more people do so, then "commons management would be deliberated through local, state, interstate, regional, and global stakeholder discussions"—ultimately leading to systems of "global constitutional governance" that favor the commons (Quilligan, 2012). However, an early 2008–2009 effort by activists to create a Coalition for the Global Commons foundered, and no new formal grand movement has emerged since.

Meanwhile, in contrast to big-science proponents of the global commons, few leftist civil-society actors are willing to envisage cooperating with government, finance, and business actors. Yet many social activists do want to see shifts to network forms of global governance—network-based governance systems—for they know that uncertainties about global governance mean difficulties for protecting and preserving the global commons. They want to adopt network-management principles designed for local commons and scale them up for application at the global level (Cogolati and Wouters, 2018). Indeed, it is encouraging for us to see Bauwens remark that "Right now, the nation-state is no longer a key instrument of change, so we must focus on building transnational open source communities of collective intelligence, i.e. a noopolitik for the noosphere" (Bauwens, 2018).

Military Perspectives on the Global Commons

The military idea of a commons is largely, but not originally, American. It focused on the sea—both early on, when Dutch jurist Hugo Grotius authored *The Freedom of the Seas, or the Right Which Belongs to the Dutch to Take Part in the East Indian Trade* in 1608 (Grotius, 1916), and later, when American naval strategist Alfred Thayer Mahan wrote in his landmark book *The Influence of Sea Power Upon History, 1660–1783* (1890) about the sea as "a great highway; or better, perhaps, of a wide common, over which men may pass in all directions" (Mahan, 1999, p. 78).

The ensuing construct, "command of the sea," expanded over time, with the identification and inclusion of air and other domains, into "command of the commons"—the construct that prevailed during the mid- to late-20th century. The term *global commons*—and its corollary, *command of the global commons*—became prominent in later U.S. military thought, notably with the 2008 *National Defense Strategy* and especially the 2010 *Quadrennial Defense Review Report* during the Obama administration.

In the U.S. view, the global commons contain four military domains: sea (or maritime), air, space, and cyber (five if land were added by counting Antarctica). What makes them part of the global commons is that they are "areas that belong to no one state and that provide access to much of the globe" (Posen, 2003, p. 8). Because no single entity owns or controls them, they become assets that lie beyond direct government control. Of these military commons, access to and use of the sea domain have been crucial for centuries, air for a century, outer space for about six decades, and cyberspace for about three decades. (See Barrett et al., 2011, p. xvi; Denmark and Mulvenon, 2010; Jasper, 2010; Jasper, 2013; Posen, 2003.)

The *global commons* is thus a multidomain concept, and military strategists prefer to view these domains as a "a complex, interactive system" (Redden and Hughes, 2011, p. 65). Though not exactly an integrated system, the four domains are so interconnected and interdependent that, operationally, they function as a whole, not just as an assemblage of parts. Together, they form the "connective tissue" (Brimley, 2010) and "sinews" (Lalwani and Shifrinson, 2013) that spread around the world; their "value lies in their accessibility, commonality, and ubiquity as a system of systems" (Barrett et al., 2011, p. 46). Indeed, a weakness or loss in any one domain (say, cyberspace) might jeopardize operations in another (say, for an aircraft carrier at sea). Accordingly, "the global commons only functions effectively because each aspect is utilized simultaneously" (Denmark and Mulvenon, 2010, p. 9). With a few word changes, this is similar to how environmental scientists and civil-society activists view their own global commons as a complex adaptive system.

What makes these commons strategically important is that they comprise "the underlying infrastructure of the global system . . . conduits for the free flow of trade, finance, information, people, and technology" (Jasper and Giarra, 2010, p. 2). Our world is so inextricably connected across these four domains that "dependable access to the commons is the backbone of the international economy and political order, benefiting the global community in ways that few appreciate or realize" (Denmark and Mulvenon, 2010, p. 1). Therefore, the global commons should be treated as "global public goods" and "global common goods." It has even been said—perhaps as an overstatement—that "every person's fate [is] tethered to the commons" (Cronin, 2010, p. ix; see also Brimley, Flournoy, and Singh, 2008; Edelman, 2010). Moreover, among nuclear strategists, creating a "new global security commons" that would include Russia has been viewed as elemental "to creating the conditions for a world without nuclear weapons" (Goodby, 2015b, p. 65).

To advance American interests since the end of World War II and throughout the Cold War, the U.S. military devoted itself to assuring that U.S. military capabilities sufficed to keep these commons accessible and usable by all in peacetime. What began as "freedom of the seas" thus evolved into favoring freedom across all of the commons—mostly for vessels, goods, and people, but also to help spread American values and ideas about openness, freedom, and democracy around the world. U.S. strategy for the global commons thus favored inclusion, not exclusion (see Flournoy and Brimley, 2010). This accords with what Teilhard might have recommended, though it is doubtful that military strategists were ever motivated by the idea of the noosphere.

In that earlier period (1945–1991), U.S. presence in the global commons was so powerful, pervasive, and singular that military strategists commended our *primacy*, *superiority*, *dominance*, and *hegemony* as being of enormous benefit—e.g., as "the key military enabler of the U.S. global power position" (Posen, 2003, p. 8), "an important enabler of globalization" (Posen, 2007, p. 563), "intrinsic to safeguarding national territory and economic interests" (Jasper and Giarra, 2010, p. 5), and "a source of US primacy and also a global public good that supported general acceptance of the unipolar world order" (Edelman, 2010, p. 77). Indeed, most of this was true—the U.S. role in securing the commons generated new opportunities for acquiring transit rights and forward bases that expanded the U.S. ability to operate as a global power and counter the ambitions of potential adversaries.

Today, however, with the world ever more globalized, multipolar, and technological, the era of the United States as guarantor of the global commons looks increasingly compromised, even jeopardized. All four domains have become congested, competitive, and contested; now, contact in any domain often risks confrontation.

The new challenges are conceptual and political, as well as military and technological; apart from the interest of the North Atlantic Treaty Organization (NATO) in the global commons (e.g., Barrett et al., 2011; NATO Supreme Allied Command Transformation, 2011), many states—notably China and Russia—disagree with U.S. views that the global commons really exist and that the world benefits from U.S. maintenance of them. Such states have laid claims to nearby sea and air spaces, objected to treating outer space as a commons, and denied letting cyberspace be a commons. Instead, these states have asserted sovereignty over portions of one commons or another, thereby expanding their security perimeters into all domains. For them, there is too much contradiction between their concepts of state sovereignty and U.S. and others' concepts of global commons (Barabanov and Savorskaya, 2018, p. 24).

In particular, China's ambitious plans to extend its political, economic, and military reach abroad, notably via its Belt and Road Initiative, seem sure to create problems in all four domains of the global commons, alarming India above all. Other new challenges come from armed nonstate actors—high-seas pirates, smugglers, and terrorists. Meanwhile, almost all actors, state and nonstate, are strengthening their capacities for access-and-area denial by acquiring advanced weapons and communications systems—a lesson they have learned from watching recent conflicts and seeing how much U.S. power projection has depended on its dominant access to and use of the global commons (Denmark and Mulvenon, 2010, p. 15; see also Brimley, 2010).

No wonder lawfare expert Craig Allen cautioned a decade ago "that an aggressive command of the commons posture may backfire and motivate other States to undertake measures to reduce the would-be commander's access or transit rights," for "claims to a 'command of the commons' seem unnecessarily provocative" (Allen, 2007, pp. 35, 38). No wonder, too, that defense expert Andrew Krepinevich concluded that "traditional means and methods of projecting power and accessing the global commons are growing increasingly obsolete—becoming 'wasting assets'" (Krepinevich, 2009, p. 18). Or that defense analyst Patrick Cronin wrote that "Securing freedom in the global commons may be the signal security challenge of the twenty-first century" (Cronin, 2010, p. ix). And no wonder that former U.S. Secretary of State George Shultz warned recently, as he has for years, of a looming "breakdown of the global commons"—for "that commons is now at risk everywhere, and in many places it no longer really exists" (Shultz, 2016).

Thus, even though U.S. military strategists might have wished to continue exercising, if not imposing, a unilateral leading role in the global commons, the time for that appears to be passing. An uncertain new era is emerging (Freeman, 2016; Murphy, 2010). Many strategists, across the political spectrum, still recognize the value of the global commons for U.S. power and influence (Ashford et al., 2019; Brands and Edelman, 2017; Bostrom, 2019; Drezner, 2019; Rose, 2019). But they also increasingly see that new conceptual and organizational approaches are needed to protect and preserve the value of the global commons. As one report put it, in the heyday of such analysis during the Obama administration,

> These trends are . . . harbingers of a future strategic environment in which America's role as an arbiter or guarantor of stability within the global commons will become increasingly complicated and contested. If this assessment is true, then a foundational assumption on which every post–Cold War national security strategy has rested—uncontested access to and stability within the global commons—will begin to erode. (Flournoy and Brimley, 2010)

Against this background, the newer analyses of how to preserve and protect the global commons to the strategic benefit of the United States and its friends and allies now mostly conclude with calls for negotiating the creation of new multilateral governance regimes, international agreements, and norms of behavior to assure the openness of the commons. Most analysts would prefer that these efforts reflect U.S. leadership, for it is a widely held view that "America must take a leadership role to ensure that access to the global commons remains a public good" (Brimley, Flournoy, and Singh, 2008, p. 15). But the United States is not in a position to impose such regimes, nor would it want to use hard power to do so—that would undermine the very notion of a global commons and its potential as a basic precept of noopolitik. Showing leadership has become a matter of having to share responsibility and work with allies and partners, in diplomatic soft-power ways akin to noopolitik—not an easy undertaking in the current environment.

The challenge is that efforts to establish governance regimes for the global commons involve not only other countries' militaries (e.g., NATO) but also various public and private actors, which can result in complex network-coordination and -cooperation problems. As Jasper and Giarra once observed,

> It is misleading to conceptualize or deal with the interests of stakeholders in the global commons independently, that is, to differentiate between the military, civil, or commercial spheres, or to segregate military service roles. This is because the domains of the commons are inherently interwoven—military maritime, space, aerospace, and cyberspace operations overlap with civilian and commercial

activities—and because the networks that enable operations or activities in the various overlapping sectors are themselves threaded together. (Jasper and Giarra, 2010, p. 3)

Denmark and Mulvenon further clarify the international challenge by concluding that

. . . the United States should renew its commitment to the global commons by pursuing three mutually supporting objectives:

- **Build global regimes:** America should work with the international community, including allies, friends, and potential adversaries, to develop international agreements and regimes that preserve the openness of the global commons.
- **Engage pivotal actors:** The United States should identify and build capacities of states and non-state actors that have the will and ability to responsibly protect and sustain the openness of the global commons.
- **Re-shape American hard power to defend the contested commons:** The Pentagon should develop capabilities to defend and sustain the global commons, preserve its military freedom of action in commons that are contested, and cultivate capabilities that will enable effective military operations when a commons is unusable or inaccessible. (Denmark and Mulvenon, 2010, p. 2, emphasis in original; see also Lalwani and Shifrinson, 2013)

Their first two recommendations are generally found in civilian pro-commons circles as well. Variants of the third point also appear in civilian circles, but without the reference to reshaping hard power—unless that reshaping were interpreted to mean a conversion into soft-power measures, a conversion that would fit with noopolitik.

Some organizational challenges are especially serious for the United States. Several reports from 2010–2012 advise strategists and planners to revamp how they approach the global commons. Accordingly, "domain-centric stovepipes have made it difficult to gain a broader, more strategic view of what changing dynamics in the sea, air, space, and cyberspace domains portend for the protection and pursuit of U.S. interests" (Brimley, 2010, pp. 121–122). One report urges reform of the "decentralized system of responsibility, in which dozens of agencies and departments are charged with securing specific aspects of the air commons" (Denmark and Mulvenon, 2010, p. 23). Another proposes to "depart from the domain-centric mindset" and "employ a holistic approach that breaks down domain stovepipes and treats the global commons not as a set of distinct geographies, but rather as a complex, interactive system" (Redden and Hughes, 2011, p. 65). Yet another calls for overcoming "inadequate governance, insufficient norms and regulations, a lack of verification measures to ensure compliance, and more often than not ineffective mechanisms for enforcement" (Barrett et al., 2011, p. xvii). Moreover, a RAND paper notes that "three global issues—climate change, water scarcity, and pandemics—[could] be posed as national security challenges with interconnected threats to the global 'commons,'" thereby making the organizational challenge even broader (Treverton, Nemeth, and Srinivasan, 2012, p. xi).

All of this makes for a daunting set of conceptual, organizational, and operational challenges. We have found no indications that these challenges have been mastered. Matters have been problematic for at least the past decade, in that "Washington has yet to articulate a diplomatic strategy to sustain access to the commons" (Denmark, 2010, p. 166).

Most recently, the inclinations of the Trump administration toward the global commons concept have been far from clear; they are still unfolding. Yet the concept remained at least partially alive in military circles as, in the waning months of the Obama administration, the Pentagon superseded its Air-Sea Battle (ASB) concept with the Joint Concept for Access and Maneuver in the Global Commons (JAM-GC), enshrining the concept in the title (Hutchens et al., 2017). Whereas ASB focused on defeating an adversary's anti-access/area-denial capabilities, JAM-GC proposed a broader approach—a "unifying framework"—for assuring freedom of action across all five warfighting domains, including land (Hutchens et al., 2017, p. 139).

Accordingly, "JAM-GC acknowledges that 'access' to the global commons is vital to U.S. national interests, both as an end in itself and as a means to projecting military force into hostile territory" (Hutchens et al., 2017, p. 137). Moreover, JAM-GC recognized that "other elements of national power—that is, a whole-of-government and coalition approach—including diplomatic, information, military, economic, financial, intelligence, and law enforcement should also be well integrated with joint force operations" (p. 138). This document is supposed to help determine strategy and doctrine for the rest of this decade and on into the next (pp. 137–139).

Moreover, the 2016 Joint Operating Environment (JOE) report, *Joint Operating Environment (JOE) 2035: The Joint Force in a Contested and Disordered World*, looking ahead to 2035, foresees a "disrupted global commons" with conflict and competition across all domains (DoD, 2016). Like previous JOEs, it acknowledged the pivotal value of the global commons for economic, military, and other matters. It also warned, "In 2035, the United States will find itself challenged in parts of the global commons as states and some non-state actors assert their own rules and norms within them" (p. 30). Indeed, "The next two decades will see adversaries building the capacity to control approaches to their homelands through the commons, and later, translating command of the nearby commons into the connective architecture for their own power projection capabilities" (p. 33).

These forecasts, written at the close of the Obama administration, were the last to refer specifically to the global commons. With the advent of the Trump administration, the *Summary of the 2018 National Defense Strategy* (DoD, 2018) never mentions *global commons* per se, referring only to "common domains" in a few spots. "Ensuring common domains remain open and free" is in the list of defense objectives (DoD,

2018, p. 4). And—to Beijing's subsequent rebuke—the document states that "We will strengthen our alliances and partnerships in the Indo-Pacific to a networked security architecture capable of deterring aggression, maintaining stability, and ensuring free access to common domains" (p. 9). At the least, military interest in the global commons concept lingers here by implication. However, another pertinent military document, *The Operational Environment and the Changing Character of Future Warfare*, which assesses matters out to the year 2050, stresses the likelihood of multidomain conflicts but never mentions the concept of the global commons (U.S. Army Training and Doctrine Command [TRADOC G-2], 2017).

Meanwhile, in nonmilitary circles, the current administration and its civilian policymakers and strategists have exhibited little interest in the global commons. To the contrary, National Space Council director Scott Pace disparaged the concept in December 2017 in harsh, dismissive terms that challenged its validity and legality:

> Finally, many of you have heard me say this before, but it bears repeating: outer space is not a 'global commons,' not the 'common heritage of mankind,' not 'res communis,' nor is it a public good. These concepts are not part of the Outer Space Treaty, and the United States has consistently taken the position that these ideas do not describe the legal status of outer space. To quote again from a U.S. statement at the 2017 COPUOS [Committee on the Peaceful Uses of Outer Space] Legal Subcommittee, reference to these concepts is more distracting than it is helpful. To unlock the promise of space, to expand the economic sphere of human activity beyond the Earth, requires that we not constrain ourselves with legal constructs that do not apply to space. (Pace, 2017)

And then, a few months later, in April 2018, Congress approved H.R. 2809—the American Space Commerce Free Enterprise Act—with a clause stating, "Notwithstanding any other provision of law, outer space shall not be considered a global commons" (U.S. House of Representatives, 2018). That represents quite a strike against the concept. In a further apparent strike, in June 2018, Federal Communications Commission director Ajit Pai repealed net neutrality rules that had kept the internet wide open, preventing internet service providers from limiting or otherwise controlling people's access—a repeal that appeared to move away from treating cyberspace as a global commons.

Will the Trump administration, or any succeeding administration, for that matter, extend anti-commons postures to the other domains—even though prior administrations, both Republican and Democratic, have long favored America's roles as primary upholder of the global commons?

It is too soon to tell. But if there is a reversion to nationalist and neo-mercantilist approaches for taking hold of territories and resources in all four domains, along with a denial that the global-commons concept has validity or legality—then this will likely lead to the alienation of pro-commons forces in environmental-science and civil-society circles, as well as a further downturn in U.S. relations with allies and partners, not to mention new difficulties with such competitors as China as it expands its global reach into all domains.

Beijing has never accepted the global-commons concept. The Trump administration's 2018–2019 call for a free and open Indo-Pacific (FOIP) might appear to favor the concept—the FOIP initiative briefly refers to the importance of ensuring free and open access to "the air, sea, land, space, and cyber commons that form the current global system," including by challenging "excessive maritime claims asserted by allies and partners, as well as those made by potential adversaries and competitors" (DoD, 2019, pp. 2, 43). However, it is not clear that protecting and preserving the global commons is much of a motivation, or interest, behind the FOIP initiative. The initiative seems more likely to destabilize matters (Swaine, 2018), to the detriment of the concept.

So, we likely have to remain patient about our hopes that attention to the global commons will favor a turn toward noopolitik anytime soon. U.S. military perspectives on the global commons have evolved in directions that seem positive for the noosphere and noopolitik. But a worldwide policy shift—by the United States, China, Russia, and others—toward neo-mercantilism and unilateralism would interrupt the long historical progression toward supporting freedom of the seas and securing the global commons. It would mean an inadvisable return to realpolitik and induce a further decline in the capacity for conflict resolution via public diplomacy and multilateral engagement. Given the huge influence that American actions have on the course of world affairs, it seems that, in the wake of a reversion in U.S. policy toward a realpolitik view of the commons, we would have to put our hopes for the noosphere and noopolitik on hold for quite a few more years.

One consideration might alter this: climate change. The U.S. military and civilian leadership in DoD have assessed that climate change is real and that it has potentially threatening security and military implications for the global commons, not to mention for other matters. Climate change has been deemed a "threat multiplier" and "an accelerant of instability or conflict" (La Shier and Stanish, 2017, passim). Key concerns include how climate change could affect not only the military's own operations, infrastructures, communities, supply chains, and budgets but also its outreach roles in humanitarian assistance, disaster relief, and border security missions, especially in the event of massive population displacements—roles that could require enhanced abilities to monitor, access, and use all of the commons quickly and efficiently (La Shier and Stanish, 2017). If climate change concerns were to help revalidate the global-commons concept, they might also confirm the need for greater contact and coordination between military strategists and civilian activists who favor that concept.

Intersecting Implications—A New Combination of Forces for the Future?

Comparing views in the civilian and military circles about the global commons illuminates significant overlaps and intersections:

- All of their definitions of *global commons* overlap—essentially, material and immaterial terrains and/or resources, located outside national jurisdictions, that amount to global public goods, thus available for mutual sharing and governance.
- All view the global commons as a set of interconnected, interdependent domains that, together, comprise a complex interactive system of systems that girds Earth.
- All have crucial interests in protecting and preserving the global commons—some for humanity's sake, others more for security's sake. All are aware that the global commons are under increasing pressures, even threats, as a result of people's behaviors.
- All believe that current governance regimes are inadequate. Work is urgently needed to create new global governance regimes, associations, and frameworks that will organize multilateral cooperation in myriad senses—i.e., intergovernmental, state-nonstate, public-private, IGO-NGO, civil-military, local-global, and hierarchical-networked—for purposes that include mutual stewardship and shared responsibility.
- All regard the global commons as strategic resources or assets, essential factors for humanity's future, around which grand strategies should be formulated. For both military and civilian actors, approaches based on soft power, not hard power, are deemed the way to pursue whatever grand strategy is proposed—i.e., noopolitik, not realpolitik.
- Global commons proponents seem to agree that another point merits greater attention: the need for sensory systems to detect and monitor what is transpiring throughout the global commons. This point is not missing from current discussions, but it is rarely highlighted, especially compared with the attention devoted to organizational matters. Yet the two matters are related—both networked sensor arrays and sensory organizations seem urgently needed for both social and scientific monitoring, including to support humanitarian assistance and disaster relief missions.

In addition to these overlaps and intersections, two significant differences stand out between civilian and military intentions toward the global commons:

- The military's intentions are focused on domain security matters; they say nothing, or very little, about societal matters. In contrast, the civilian circles do aim to transform societies so that they become better designed to live with, and benefit from, the global commons. The big-environmental-science circle has made myriad proposals for social, economic, and political reforms, some quite radical. The leftist social-activist civil-society circle foresees societies being radically transformed as a result of pro-commons forces.

- Both military and civilian proponents of the global commons talk about the importance of hegemony—but in opposite ways. The military has aspired to hegemonic control over the global commons. In contrast, an oft-mentioned goal of civil-society commoners is *counter-hegemonic* power—making pro-commons forces so strong that they can counter the hegemonic power of today's public and private sectors, indeed of capitalism itself. This makes it difficult to imagine today's pro-commons social activists relating well to today's pro-commons military strategists. But the day might come, especially if and as climate change and its effects are perceived to be an overriding mutual concern.

These findings support our sense that the noosphere and noopolitik concepts will fare better in the future the more they are associated with the global commons concept. And the latter will flourish the more it is associated with the noosphere and noopolitik. This seems likely because both the global commons and the noosphere have links to the biosphere. Recognizing the noosphere's links to the global commons might then help put noopolitik back on track in various strategic issue areas.

True as that might be, optimism and enthusiasm are barely warranted right now. Looking ahead with the current political environment in mind—especially the orientations of today's leaders in Washington, Beijing, and Moscow—what seems most in need of protection and preservation are not just the global commons per se, but the very concept of global commons itself. An administration in Washington that disparages this long-standing strategic concept will too easily play into the hands of Beijing and Moscow. Neither of them has ever accepted the concept; both would rather pursue their own grand strategies without it. Leadership on behalf of the global commons— and prospects for the noosphere and noopolitik—would then fall more than ever to the circles of nonstate actors that we highlighted earlier.

CHAPTER EIGHT

Getting Back on Track Through Noopolitik

At this fraught strategic moment, when it is advisable, perhaps necessary, for U.S. strategy and diplomacy to lead the international community in the direction of noopolitik, conditions are once again not ripe for doing so. Indeed, it might be a while before propitious conditions reemerge, for as America's soft power rises and falls, so do the prospects for the global diffusion of noopolitik. And right now, American soft-power capabilities are very much in question.

The United States has long stood for cherished ideals—freedom, equality, opportunity. It has also stood for ethical ways of doing things: competing openly and fairly, working in concert with partners, seeking the common good, respecting others' rights and responsibilities, and resorting to warfare only after exhausting nonmilitary options. By doing so, the United States built its legitimacy and credibility as a global power in the 20th century. Lately, however, leaders and publics around the world have become increasingly doubtful that Americans are as deeply dedicated to the ideals and practices that they have traditionally championed. As a result, U.S. public diplomacy is on the defensive more than ever. Oddly, China is sometimes said to be more effective at soft-power appeals and techniques.

What would reinvigorate the prospects for noopolitik? Renewal of a clear intent to favor nonmilitary strategies, operate in partnerships, and abide by stringent ethical standards would surely help. So would further advancing the global commons. Yet, specifics aside, the key might well be a revitalization of a deep sense that ideas matter, along with a better grasp of how ideas, expressed as narratives, move people to think and act in strategic ways—more along the lines of the complex efforts made during the Cold War than the simplifications seen in recent decades. Strings of overseas conflicts and other events, including domestic troubles, have undermined preferred American narratives about fostering a peaceful, prosperous, civilized, democratic world in which all nations are bound together by shared norms and values.

Look around: U.S. hard-power approaches to one conflict after another seem only to have incurred high costs and greater risks, in return for scarcely discernible benefits. Hard-power efforts have failed to overthrow Bashar al-Assad's rule in Syria, have fostered Iranian influence over Iraq, and have perpetuated a quagmire in Afghanistan. Little has been done to impress (much less impose) U.S. will on China and

Russia, not to mention North Korea and Venezuela. Realpolitik by itself, in either its military or economic applications, holds no real promise of solving these conflicts and other challenges.

It is well past time to invigorate the application of noopolitik. Whereas realpolitik is typically about whose military or economy wins, noopolitik is ultimately about whose story wins. Thus, it is about affecting cognitions of all kinds—inspiring, attracting, persuading, convincing, sharing, and just plain listening, as well as disapproving, dissuading, cajoling, and maybe even shunning at times. It means communicating and collaborating with partners and allies, both state and nonstate actors, and seeking them out rather than going it alone or insisting on a singular primacy. This requires the careful design and deployment of strategic narratives and messages for finding common ground around a common good; imparting cautionary ideas about where a society's evolution is headed; shaping people's social space-time-agency perceptions; framing key values and sharing best practices; or letting someone know "there's a better way." The list goes on.

And it is a list of ends, ways, and means that, above all, require diplomacy, often especially public diplomacy, along with cultured leadership. Noopolitik is far more a diplomatic than a military or intelligence enterprise. And, like realpolitik, its effectiveness depends on the presence of skilled strategists, strong agencies, and related apparatuses and capabilities. Unfortunately, much of what used to exist along these lines, much of it designed to win the Cold War, has been dismantled and devalued—by administrations from both major political parties—since we started formulating noopolitik over twenty years ago. Indeed, the Trump administration, with its penchant for relying on hard power and "deal power," has so far done little to rebuild the institutional and other preconditions for using noopolitik effectively.[1] Although this need to rebuild is an implication of our work, its urgency is more fully revealed in the writings of others (e.g., Albright, 2018; Burns, 2019a; Burns, 2019b; Dobbins, 2019; Farrow, 2018; Pandith, 2019; Walt, 2019).

The Way Ahead as We Previously Saw It

We have expected, as noted in our 1999 and 2007 writings, that strategists and diplomats would feel challenged to focus on how best to develop the noosphere and conduct noopolitik. Much as the rise of realpolitik depended on the development and exploitation of the geosphere (whose natural resources enhance state power), so will the rise of noopolitik depend on the development of the noosphere. The two go hand in hand. To pursue this, measures will have to be identified that, in addition to fostering the rise of

---

[1]   Counterarguments explaining, and praising, the Trump administration's approach to foreign policy and strategy appear in Colby and Mitchell, 2020; and Navarro, 2019.

a noosphere, are geared to facilitating the effectiveness of soft power, the deepening of global interconnections, the strengthening of transnational civil-society actors, and the creation of conditions for governments to be better able to act conjointly by seeking cooperative advantages with both state and nonstate actors.

Our first writing on this topic (1999) noted some measures for U.S. policy and strategy that could assist with the development of the noosphere and noopolitik. All were drawn from discussions back then about issues raised by the advance of the information revolution, and we thought that strategists and diplomats would be well advised to take an interest in them. These measures included the following:

- supporting the expansion of cyberspace connectivity around the world, including where this runs counter to the preferences of authoritarian regimes
- promoting freedom of information and communications as a worldwide right
- developing multitiered information-sharing systems, not only to ensure cyberspace safety and security, but also to create shared infospheres for openly addressing other matters
- creating "special media forces" that could be dispatched into conflict zones to help settle disputes through the discovery and dissemination of accurate information
- opening diplomacy to greater coordination between state and nonstate actors, especially NGOs (Arquilla and Ronfeldt, 1999; see also Ronfeldt and Arquilla, 2007).

These remain pertinent measures. Ultimately, developing the noosphere and noopolitik will involve more than just asserting, sharing, and instituting the particular values, norms, ethics, laws, and other aspects of soft power that an actor aims to uphold. Specific policies, strategies, mechanisms, and infrastructures will have to be elaborated that make noopolitik significantly different from and more effective than realpolitik for dealing with issues that may range from promoting democracy to pressuring regimes, such as those in Iran and North Korea, as well as resolving global environmental and human rights issues. Skilled diplomats and strategists are bound to face choices as to when it is better to emphasize realpolitik or noopolitik, alternate between them, or apply hybrid courses of action, especially when dealing with a recalcitrant adversary who has been able to resist realpolitik types of pressures and is learning to apply dark forms of noopolitik against the United States.

As an urgent reason to revive the prospects for noopolitik, we noted back then that several worldwide wars of ideas were underway. The most evident had spiritual, religious, ideological, philosophical, and cultural aspects and were largely taking place on the internet. In such wars of ideas, we further remarked, one's information posture matters as much as one's military posture. And, at that time, the U.S. information posture did not appear to be well designed, nor was it even regarded as a "posture."

Everything discussed so far comes from what we concluded about noopolitik and its prospects back in 1999 and 2007. All of our points look as valid now as they did back then.

Scattered Signs of Slow Progress in the Interim

While preparing this piece, we found that progress has occurred regarding some of the observations and recommendations we fielded in our 1999 and 2007 studies. Other strategists also made gains in generally recognizing the challenges we face. But altogether, it has been slow, tentative, scattered progress, at best.

In one good sign, a recent RAND report states that "The United States . . . needs an updated framework for organizing its thinking about the manipulation of infospheres by foreign powers determined to gain competitive advantage" (Mazarr, Casey, et al., 2019, p. xii). Indeed, "the United States and other democratic countries must begin to think more strategically about the information environment, their vulnerabilities, and also potential advantages" (p. 5). We agree.

Also positive is that our 1999 idea to create "special media forces" might be gaining support. Another recent RAND report about how to counter political warfare includes the following recommendation:

> Recommendation 8: DoD and State should support deployment of special operations forces in priority areas deemed vulnerable to political warfare threats as an early and persistent presence to provide assessments and develop timely and viable options for countering measures short of conventional war. Interviews with SOF [special operations forces] have stressed this requirement above all others. (Robinson et al., 2018, p. 315)

In similar veins, Daniel Gouré, 2017, proposed the creation of rapid-reaction information cells to track and counter Russian disinformation operations, and Peter Pomerantsev, 2019, proposed the creation of "a new set of civil society actors who combine the values of accurate media with engagement skills and an understanding of how propagandists prey on polarization, inflaming divides." One of the most advanced models for this might be in Estonia, which has developed a volunteer cyber army that serves much like a digital national guard (Gross, 2018). In the words of journalist Christa Case Bryant, the volunteers of the Cyber Defense League "are the modern-day minutemen of Estonia. They are part of a broader electronic bulwark this Baltic nation has built" (Bryant, 2020).

Other recent RAND analyses that resonate with our ideas for noopolitik include proposals to create a Center for Cognitive Security to assess the shifting information environment (Waltzman, 2017), as well as to create a National Political Warfare Center to pursue an all-of-government approach to political warfare (Cleveland et al., 2018;

see also Robinson et al., 2018). Both are interesting ideas. Meanwhile, at the Naval Postgraduate School, a curriculum on "Information Strategy and Political Warfare" speaks to many of the issues we have raised.

Elsewhere, analysts at the Atlantic Council have observed that "The United States has fallen behind the EU [European Union], both in terms of conceptual framing and calls for concrete actions to meet the disinformation challenge" (Polyakova and Fried, 2019, p. 7). They have called for the U.S. government to establish an interagency Counter-Disinformation Center and to help create a transatlantic Counter-Disinformation Coalition (p. 16).

Furthermore, analysts at Harvard's Belfer Center, after noting the increasing significance of "information power" and "information geopolitics," have urged Washington to develop a well-coordinated "national information strategy" (Rosenbach and Mansted, 2019, p. 14). In particular, they propose creation of a federally funded research and development center (FFRDC), one of whose purposes would be to "craft and promote genuine narratives" (p. 20). Done right, this would protect against "information mercantilism" (p. 7 ff.) and "information authoritarians" (p. 11 ff.). Yet another potentially positive development is the Weaponized Narrative Initiative of the Center on the Future of War (Allenby and Garreau, 2017).

Meanwhile, inside the U.S. government, an old effort to advance the concept of "information engagement" during 2008–2009 has, for the present, been superseded by a new idea for "information statecraft" (Trump, 2017, p. 3) that appears to have foundered. But both initiatives pointed out that major internal organizational changes are needed to improve U.S. information strategy—a point that has been made for decades, but at least it is still being made.

One place where such changes are occurring is inside the U.S. Department of State. Public diplomacy is being elevated as a priority, including the matter of "information engagement," which still survives there. The Under Secretary for Public Diplomacy and Public Affairs now has a new Bureau of Global Public Affairs alongside its Global Engagement Center ("State Department Officials on Public Diplomacy," 2019). Both the bureau and the center are mainly focused on extending the State Department's social media presence and combatting disinformation and other issues raised by current information-warfare problems. Hopefully, this area within the State Department will evolve in the direction of noopolitik, but a wait-and-see stance might be advisable for now.

Even so, in what might be a questionable move for public diplomacy, the White House reportedly is inclined to call for greater private-sector involvement, for example by cutting and rebranding old legacy platforms, such as Voice of America. Moreover, ideas to recreate the U.S. Information Agency have gained little traction. This once-valuable instrument for public diplomacy during the Cold War was closed in 1999, with pieces of it redistributed across multiple agencies—most of them allotted to the State Department. In some views (e.g., Roa, 2017), it should be revived and redirected to assist with today's information-warfare campaigns. Perhaps this might extend to recreating something like the renowned Active Measures Working Group that existed during the Reagan administration to help construct strategic communications and counter-narratives for ending the Cold War (for history, see Schoen and Lamb, 2012).

Concern about these matters has grown stronger in the military as well. Of the numerous signs of this, one that catches our eye is a 2017 study mentioned earlier, *The Operational Environment and the Changing Character of Future Warfare*. A key point it makes is that "the physical dimension of warfare may become less important than the cognitive and the moral. Military operations will increasingly be aimed at utilizing the cognitive and moral dimensions to target an enemy's will" (U.S. Army Training and Doctrine Command [TRADOC G-2], 2017, p. 18). This is quite a contrast to trends inside DoD a few years earlier. In 2011, DARPA initiated a new program, Social Media in Strategic Communication, to focus on preempting and preparing for online propaganda battles. But before long, as Renée DiResta writes, "The premise was ridiculed as an implausible threat, and the program was shut down in 2015" (DiResta, 2018a). She adds, quite critically, that "Now, both governments and tech platforms are scrambling for a response. The trouble is that much of the response is focused on piecemeal responses to the last set of tactics." Even so, at least a realization was finally dawning that conflict was shifting "from targeting infrastructure to targeting the minds of civilians" (DiResta, 2018a).

All of these interim developments, though scattered and halting, look positive for the future of noopolitik. Yet, overall, noopolitik still seems to be an idea for the future. Traditional power politics—realpolitik—has provided the main basis for U.S. foreign policy and strategy in the decades since 9/11. By now, various new wars of ideas are underway, and the U.S. government is still not participating in them in ways reflective of the noopolitik paradigm. Washington's continued threats of military force and coercive diplomacy imply the persistent primacy of older—and ever less effective—forms of statecraft. Moreover, this retrogressive American view has had an inhibiting effect on allies and friends, while at the same time reaffirming the realpolitik proclivities of rivals and competitors. Despite all of this, we still remain optimistic about the long-term promise of noopolitik.

New Information Strategy Measures for the Way Ahead

All of the aforementioned leads to several new recommendations for policy and strategy, in addition to the ones listed earlier. We believe these new findings are worth emphasizing about the prospects for the noosphere and noopolitik:

- **Back to basics—rethink "soft power":** We should not have to list this problem—it should be cleared up by now—but it is not. Soft power remains a compelling concept. But, as we discussed in Chapter Six, its definition has been biased since its upbeat coinage several decades ago; its potential dark sides have received far too little attention. We tried to warn over a decade ago that dark uses of soft power were "perhaps the most serious problem confronting American public diplomacy today" (Ronfeldt and Arquilla, 2007). The recent study about "sharp power" (Walker and Ludwig, 2017b) adds credence to our call for rethinking soft power to include recognition of its dark sides, including its uses for deception, disinformation, distortion, distraction, and other elements of cognitive and political warfare. Such a rethinking should seek to achieve not only a better definition of soft power but also a better understanding of noopolitik, its purposes, and its uses as part of America's conceptual arsenal.

- **Uphold "guarded openness" as a guiding principle:** This deliberately ambivalent word pairing means remaining open politically, economically, culturally, and even militarily (with allies, especially), in accordance with liberal values, while also voicing criteria and creating mechanisms for guardedness as a filtering system to allay risks inherent in pursuing open information strategies. Achieving an optimal balance between guardedness and openness might prove nettlesome in some issue areas—Moscow, Beijing, and other authoritarian regimes are skilled at exploiting openness—but the principle of "guarded openness" expresses very well how a democracy should behave. We proposed it two decades ago (Arquilla and Ronfeldt, 1997; Arquilla and Ronfeldt, 1998); it seems more salient now—not only for information strategy but also for a broader array of American and allied foreign-policy concerns that involve global flows (e.g., trade, immigration, technology). Noopolitik can only benefit from upholding a general principle favoring openness.

- **Take up the cause of the "global commons" as a pivotal issue area:** As we found in the preceding section, this has been a pivotal issue area for civilian activists and military strategists, though it has yet to receive widespread public recognition and is lately under challenge. We have long listed democracy promotion, human rights, the environment, and conflict resolution as issue areas that would benefit from skillful applications of noopolitik, but they were easy to list because they are so much in the public eye. The global commons rarely is. But now we see that preserving, protecting, and promoting the global commons concept—the pursuit of a secure, sustainable global commons—could be a vital addition to our list. The prospects for noopolitik depend on the prospects for the noosphere, and the future of the noosphere appears to depend on the future of the global commons—a progression in which the one cannot evolve properly without the other.

- **Institute a formal requirement for periodic reviews of the American "information posture":** We first mentioned this idea in 2007. It deserves elaboration and attention—and emulation by other nations. The U.S. military posture receives regular assessments. So do aspects of America's economic posture (even though it is not exactly called that). Information is now of such strategic importance that methodologies and measures should be deliberately designed for assessing one's information posture globally. The creation of a new interagency office, even an FFRDC, might be advisable to accomplish this and to draw out the full variety of implications for policy and strategy, say in the form of a periodic National Information Strategy document. This could be of great benefit for conducting noopolitik, as well as for understanding the status of the noosphere and the specific stocks and flows of information that comprise it. The United States has a pressing need for such an undertaking, but the fundamental concept of an "information posture" might have broad international application as well.

Such measures, combined with our earlier recommendations—not to mention some of the proposals we mentioned from other sources—would go a long way toward helping advance noopolitik as an approach to information strategy, putting it on a more even plane with the abiding appeal of realpolitik.

Coda . . . for Now

As noopolitik (hopefully) diffuses in the years and decades ahead, strategists will gradually figure out how different it is from realpolitik. Noopolitik requires a fresh way of looking at the world—a new mindset, a different knowledge base, new assessment methodologies. How to look at hard power, and thus realpolitik, is quite standardized by now. But how best to understand and use soft power is far from settled. Noopolitik depends on knowing—and finding new ways of knowing—about ideational, cognitive, and cultural matters that have not figured much in traditional statecraft.

As we have always said, noopolitik is ultimately about whose story wins. Thus, the kinds of stories, or narratives, that matter in noopolitik must be carefully constructed to suit the context. That narratives are crucial for maneuvering in today's world is widely accepted; but designing them remains more of an art than a science, and there is still plenty of room for new ideas about how to build expertise.[2] Indeed, graduate-level courses to teach grand strategy might have to be thoroughly revised—to be effective, noopolitik strategists might need to be steeped in theoretical matters that today receive little systematic attention, such as social evolution (Ronfeldt, 1996) and social cognition (Ronfeldt, 2018). That might be especially needed if a goal is to promote democracy better than U.S. policies have done in recent decades. Indeed, as a

---

[2]   The need for this in other fields as well is aptly articulated in Robert Shiller's recent calls for "narrative economics" (2017; 2019). Also pertinent is Amy Zalman's work on strategic narratives and strategic foresight (e.g., 2012; 2019).

graduate school philosophy professor used to tell us, paraphrasing Kurt Lewin, "there is nothing more practical than a good theory" (McCain, 2015).

Right now, however, our new proposals, plus our old ones, point to urgent needs to revitalize diplomacy, especially public diplomacy (see Snow and Cull, 2020). It remains in such decline and disarray that we feel a need to close, as we did our 1998 and 2007 studies, by emphasizing this point. Turning to noopolitik depends on having strong diplomatic capabilities, especially for wielding soft power. Without them, our leaders in Washington will keep being tempted to rely on realpolitik and its all-too-attractive penchant for hard power.

We advise them to heed the wisdom of Hans Morgenthau, the father of modern realpolitik. He warned that "there is the misconception . . . that international politics is so thoroughly evil that it is no use looking for ethical limitations of the aspirations for power" (Morgenthau, 1948, p. 175). This is why Morgenthau heralded "the increasing awareness on the part of most statesmen of certain ethical limitations restricting the use of war as an instrument of international politics" (p. 180).

In other words, by invoking ethics and ethical limitations, this iconic arch-realist showed an early inclination toward noopolitik. But it seems to have gone missing from the sensibilities of too many of today's leaders around the world. Perhaps they still do not grasp the complex implications of the information age, preferring instead to cling to mental models from simpler eras. Our hope is that the pull of the past, in particular of power politics, will yield as the promise of a better future is glimpsed through the lens of the noosphere and noopolitik.

The measures we have mentioned in this Perspective can open up transformational possibilities and opportunities for shifting away from realpolitik toward noopolitik, thereby better attuning statecraft to the information age. This could help burnish the image of the United States and its allies in the world once again, potentially lessening the bitterness and violence of conflicts, revitalizing diplomacy, especially public diplomacy, and setting the world on a course toward sustainable peace and prosperity. Whereas realpolitik treats international relations as intractably conflictual, the starting point for noopolitik is faith in upholding our common humanity, along with a belief that, in statecraft, ideas can matter more than armaments.

Around the world, intense debates are occurring among grand strategists over whether the future of the international system will revolve more around unipolar, bipolar, or multipolar forces; whether those forces will favor cooperation, competition, or conflict; whether primacy or retrenchment is the better path; and what all of this might mean for power politics—and especially for geopolitics and realpolitik. Our point is that noopolitical forces and noopolitik are going to increasingly matter, regardless of whether the future global configuration is unipolar, bipolar, or multipolar (or something else) and regardless of what kinds of relations ensue. We do not want to be so alone in sensing this.

Bibliography

Aberkane, Idriss J., "A Simple Paradigm for Noopolitics: The Geopolitics of Knowledge," *E-International Relations*, October 15, 2015. As of March 18, 2020:
https://www.e-ir.info/2015/10/15/
a-simple-paradigm-for-noopolitics-the-geopolitics-of-knowledge/

Albright, Madeleine, "Will We Stop Trump Before It's Too Late?" *New York Times*, April 6, 2018. As of March 18, 2020:
https://www.nytimes.com/2018/04/06/opinion/sunday/
trump-fascism-madeleine-albright.html

Allen, Craig H., "Command of the Commons Boasts: An Invitation to Lawfare?" *International Law Studies*, Vol. 83, No. 21, 2007, pp. 21–50.

Allen, T. S., and A. J. Moore, "Victory Without Casualties: Russia's Information Operations," *Parameters*, Vol. 48, No. 1, Spring 2018, pp. 59–71.

Allenby, Brad, and Joel Garreau, eds., *Weaponized Narrative: The New Battlespace*, Washington, D.C.: Weaponized Narrative Initiative, Center on the Future of War, March 21, 2017.

Allison, Graham, "The New Spheres of Influence: Sharing the Globe with Other Great Powers," *Foreign Affairs*, Vol. 99, No. 2, March–April 2020, pp. 30–40.

Anderson, Benedict, *Imagined Communities: Reflections on the Origin and Spread of Nationalism*, Revised Edition, New York: Verso, 1991.

Ang, Yuen Yuen, "Autocracy with Chinese Characteristics: Beijing's Behind-the-Scenes Reforms," *Foreign Affairs*, Vol. 97, No. 3., May–June 2018, pp. 39–46.

Applebaum, Anne, "The False Romance of Russia," *The Atlantic*, December 12, 2019. As of March 18, 2020:
https://www.theatlantic.com/ideas/archive/2019/12/false-romance-russia/603433/

Argüelles, José, *Manifesto for the Noosphere: The Next Stage in the Evolution of Human Consciousness*, Berkeley, Calif.: North Atlantic Books, Evolver Editions, 2011.

Arquilla, John, and David Ronfeldt, eds., *In Athena's Camp: Preparing for Conflict in the Information Age*, Santa Monica, Calif.: RAND Corporation, MR-880-OSD/RC, 1997. As of March 18, 2020:
http://www.rand.org/publications/MR/MR880/index.html

Arquilla, John, and David Ronfeldt, "Preparing for Information-Age Conflict: Part 2 Doctrinal and Strategic Dimensions," *Information, Communication & Society*, Vol. 1, No. 2, Summer 1998, pp. 121–143.

———, *The Emergence of Noopolitik: Toward An American Information Strategy*, Santa Monica, Calif: RAND Corporation, MR-1033-OSD, 1999. As of March 18, 2020:
http://www.rand.org/pubs/monograph_reports/MR1033/index.html

———, *Swarming and the Future of Conflict*, Santa Monica, Calif.: RAND Corporation, DB-311-OSD, 2000. As of March 18, 2020:
https://www.rand.org/pubs/documented_briefings/DB311.html

———, *Networks and Netwars: The Future of Terror, Crime, and Militancy*, Santa Monica, Calif.: RAND Corporation, MR-1382-OSD, 2001. As of March 18, 2020:
http://www.rand.org/pubs/monograph_reports/MR1382/index.html

Ashford, Emma, Hal Brands, Jasen J. Castillo, Kate Kizer, Rebecca Friedman Lissner, Jeremy Shapiro, and Joshua R. Itzkowitz Shifrinson, *New Voices in Grand Strategy*, Washington, D.C.: Center for a New American Security, 2019.

Atran, Scott, Hammad Sheikh, and Ángel Gómez, "For Cause and Comrade: Devoted Actors and Willingness to Fight," *Cliodynamics: The Journal of Quantitative History and Cultural Evolution*, Vol. 5, No. 1, 2014, pp. 41–57. As of March 18, 2020:
https://escholarship.org/uc/item/6n09f7gr

Bacevich, Andrew J., *Washington Rules: America's Path to Permanent War*, New York: Henry Holt and Company, 2010.

Barabanov, Oleg, and Ekaterina Savorskaya, "Global Environmental Ideologies: Can the Conflict Between Humans and Nature Be Overcome?" Moscow: Valdai Discussion Club Report, December 2018. As of March 18, 2020:
http://valdaiclub.com/a/reports/global-environmental-ideologies/

Barno, David, and Nora Bensahel, "A New Generation of Unrestricted Warfare," *War on the Rocks*, April 19, 2016. As of March 18, 2020:
https://warontherocks.com/2016/04/a-new-generation-of-unrestricted-warfare/

Barrett, Mark, Dick Bedford, Elizabeth Skinner, and Eva Vergles, *Assured Access to the Global Commons*, Norfolk, Va.: Supreme Allied Command Transformation, North Atlantic Treaty Organization, April 2011. As of March 18, 2020:
http://www.act.nato.int/images/stories/events/2010/gc/aagc_finalreport.pdf

Baue, Bill, and Ralph Thurm, "What Are Thresholds and Allocations, and Why Are They Necessary for Sustainable System Value Creation?" *Medium*, May 9, 2018. As of March 18, 2020:
https://medium.com/@r3dot0/what-are-thresholds-allocations-and-why-are-they-necessary-for-sustainable-system-value-fe127483c407

Bauwens, Michel, "P2P: A New Cycle of Post-Civilizational Development," transcript of interview with Rajani Kanth, *Commons Transition*, April 19, 2018. As of March 18, 2020:
http://commonstransition.org/18186-2/

Bauwens, Michel, Vasilis Kostakis, and Alex Pazaitis, *Peer to Peer: The Commons Manifesto*, London: University of Westminster Press, 2019.

Bauwens, Michel, Vasilis Kostakis, Stacco Troncoso, and Ann Marie Utratel, *Commons Transition and P2P: A Primer*, Transnational Institute, P2P Foundation, March 2017. As of March 18, 2020:
https://p2pfoundation.net/
wp-content/uploads/2017/09/commons_transition_and_p2p_primer_v9.pdf

Bauwens, Michel, and Jose Ramos, "Re-Imagining the Left Through an Ecology of the Commons: Toward a Post-Capitalist Commons Transition," draft, P2P Foundation Wiki, January 2018. As of March 18, 2020:
https://wiki.p2pfoundation.net/Why_We_Need_a_Post-Capitalist_Commons_Transition_Today

Benkler, Yochai, Robert Faris, and Hal Roberts, *Network Propaganda: Manipulation, Disinformation, and Radicalization in American Politics*, New York: Oxford University Press, 2018.

Berlin, Geoffrey, "National Security and Climate Change," *The Globalist*, May 30, 2013. As of March 18, 2020:
https://www.theglobalist.com/national-security-and-climate-change/

Bew, John, "The Real Origins of Realpolitik," *National Interest*, February 25, 2014. As of March 18, 2020:
https://nationalinterest.org/print/article/the-real-origins-realpolitik-9933

Bloom, Solomon F., "The 'Withering Away' of the State," *Journal of the History of Ideas*, Vol. 7, No. 1, January 1946, pp. 113–121.

Bollier, David, *The Rise of Netpolitik: How the Internet Is Changing International Politics and Diplomacy*, Washington, D.C.: Aspen Institute, Report of the Eleventh Annual Aspen Institute Roundtable on Information Technology, 2003.

———, "NATO Misconstrues the Commons," blog post, David Bollier: News and Perspectives on the Commons, December 18, 2010. As of March 18, 2020:
http://www.bollier.org/nato-misconstrues-commons

———, "The Commons as a Growing Global Movement," blog post, David Bollier: News and Perspectives on the Commons, June 14, 2014. As of March 18, 2020:
http://www.bollier.org/blog/commons-growing-global-movement

Bollier, David, and Silke Helfrich, eds., *The Wealth of the Commons: A World Beyond Market and State*, Amherst, Mass.: Levellers Press, 2012.

———, *Patterns of Commoning*, Amherst, Mass:, Common Strategies Group, 2015.

Bollier, David, and Jonathan Rowe, *The Commons Rising*, ed. Peter Barnes and Seth Zuckerman, Minneapolis, Minn.: Tomales Bay Institute, 2006.

Bossenbroek, Martin, *The Boer War*, trans. Yvette Rosenberg, New York: Seven Stories Press, 2018.

Bostrom, Nick, "The Vulnerable World Hypothesis," *Global Policy*, Vol. 10, No. 4, November 2019, pp. 455–476.

Boulding, Elise, *Building a Global Civic Culture: Education for an Interdependent World*, New York: Teachers College Press, 1988.

Brands, Hal, and Eric Edelman, "America and the Geopolitics of Upheaval," *National Interest*, June 21, 2017. As of March 18, 2020:
https://nationalinterest.org/print/feature/america-the-geopolitics-upheaval-21258

Brannen, Peter, "The Anthropocene Is a Joke," *The Atlantic*, August 13, 2019. As of March 18, 2020:
https://www.theatlantic.com/science/archive/2019/08/arrogance-anthropocene/595795/

Brimley, Shawn, "Promoting Security in Common Domains," *Washington Quarterly*, Vol. 33, No. 3, July 2010, pp. 119–132.

Brimley, Shawn, Michèle A. Flournoy, and Vikram J. Singh, *Making America Grand Again: Toward a New Grand Strategy*, Washington, D.C.: Center for a New American Security, June 2008.

Brin, David, *Earth*, New York: Bantam Books, 1990.

Bryant, Christa Case, "Cybersecurity 2020: What Estonia Knows About Thwarting Russians," *Christian Science Monitor*, February 4, 2020. As of March 18, 2020:
https://www.csmonitor.com/World/Europe/2020/0204/
Cybersecurity-2020-What-Estonia-knows-about-thwarting-Russians

Burns, William J., "How to Save the Power of Diplomacy," Carnegie Endowment for International Peace, March 8, 2019a. As of March 18, 2020:
https://carnegieendowment.org/2019/03/08/how-to-save-power-of-diplomacy-pub-78542

———, "The Lost Art of American Diplomacy: Can the State Department Be Saved?" *Foreign Affairs*, Vol. 98, No. 3, May–June 2019b, pp. 98–107.

Bush, George H. W., "Transcript of President Bush's Address on End of the Gulf War," *New York Times*, March 7, 1991.

Butter, Michael, and Peter Knight, eds., *Routledge Handbook of Conspiracy Theories*, New York: Routledge, 2020.

Campbell, Kurt M., and Jake Sullivan, "Competition Without Catastrophe: How America Can Both Challenge and Coexist with China," *Foreign Affairs*, Vol. 98, No. 5, September–October 2019, pp. 96–110.

Christian, David, "The Noösphere," Edge.org, 2017. As of March 18, 2020:
https://www.edge.org/response-detail/27068

Cleveland, Charles T., Ryan Crocker, Daniel Egel, Andrew M. Liepman, and David Maxwell, *An American Way of Political Warfare: A Proposal*, Santa Monica, Calif.: RAND Corporation, PE-304, 2018. As of March 18, 2020:
https://www.rand.org/pubs/perspectives/PE304.html

Cogolati, Samuel, and Jan Wouters, eds., *The Commons and a New Global Governance*, Cheltenham, UK: Edward Elgar Publishing, 2018.

Colby, Elbridge A., and A. Wess Mitchell, "The Age of Great-Power Competition: How the Trump Administration Refashioned American Strategy," *Foreign Affairs*, Vol. 99, No. 1, January–February 2020, pp. 118–130.

Cronin, Patrick, "Foreword," in Scott Jasper, ed., *Securing Freedom in the Global Commons*, Stanford University Press, 2010, pp. ix–xv.

Csikszentmihalyi, Mihaly, "I Must Confess to Being Perplexed," in John Brockman, ed., *Is the Internet Changing the Way You Think? The Net's Impact on Our Minds and Future*, New York: Harper Perennial, 2011, pp. 374–375.

DARPA—*See* Defense Advanced Research Projects Agency.

Defense Advanced Research Projects Agency, "Special Notice (SN) DARPA-SN-17-45 Discover DSO Day (D3)," Amendment 1, June 15, 2017. As of March 18, 2020:
https://www.darpa.mil/attachments/DARPA-SN-17-45%20Amendment%201.pdf

Denmark, Abraham M., "Managing the Global Commons," *Washington Quarterly*, Vol. 33, No. 3, July 2010, pp. 165–182.

Denmark, Abraham M., and James Mulvenon, *Contested Commons: The Future of American Power in a Multipolar World*, Washington, D.C.: Center for a New American Security, January 2010. As of March 18, 2020:
https://s3.amazonaws.com/files.cnas.org/documents/CNAS-Contested-Commons-Capstone_0.pdf

Dettling, Christopher Richard Wade, "Computational Superpower: Select Bibliography," *Medium*, November 4, 2018. As of March 18, 2020:
https://medium.com/@christopherrichardwadedettling/
world-mind-noosphere-select-bibliography-94097078f164

Diamond, Larry, "America's Silence Helps Autocrats Triumph," *Foreign Policy*, September 6, 2019. As of March 18, 2020:
https://foreignpolicy.com/2019/09/06/americas-silence-helps-autocrats-triumph-democratic-rollback-recession-larry-diamond-ill-winds/

Diamond, Larry, and Brad Carson, "Jaw-Jaw: How Chinese Sharp Power Takes Aim at American Democracy," *War on the Rocks*, podcast, February 5, 2019. As of March 18, 2020:
https://warontherocks.com/2019/02/
jaw-jaw-how-chinese-sharp-power-takes-aim-at-american-democracy/

Diamond, Larry, and Orville Schell, eds., *China's Influence and American Interests: Promoting Constructive Vigilance*, Revised Edition, Stanford, Calif.: Report of the Working Group on Chinese Influence Activities in the United States, Hoover Institution Press, 2019. As of March 18, 2020:
https://www.hoover.org/research/
chinas-influence-american-interests-promoting-constructive-vigilance

Dickens, Charles, *A Tale of Two Cities*, London: Chapman and Hall, 1859.

Dinerman, Taylor, "The End of the Soft-Power Delusion," *National Review*, December 31, 2019. As of March 18, 2020:
https://www.nationalreview.com/2019/12/american-foreign-policy-economic-military-strength-vital/

DiResta, Renée, "The Digital Maginot Line," *Ribbonfarm* blog, November 28, 2018a. As of March 18, 2020:
https://www.ribbonfarm.com/2018/11/28/the-digital-maginot-line/

———, "What We Now Know About Russian Disinformation," *New York Times*, December 17, 2018b. As of March 18, 2020:
https://www.nytimes.com/2018/12/17/opinion/russia-report-disinformation.html

Dobbins, James, "Time to Return to the Basics of Statecraft," *RAND Blog*, September 4, 2019. As of March 18, 2020:
https://www.rand.org/blog/2019/09/time-to-return-to-the-basics-of-statecraft.html

DoD—*See* U.S. Department of Defense.

Doshi, Rush, "China Steps Up Its Information War in Taiwan: Taiwan's Election Is a Test Run for Beijing's Worldwide Propaganda Strategy," *Foreign Affairs*, January 9, 2020. As of March 18, 2020:
https://www.foreignaffairs.com/articles/china/
2020-01-09/china-steps-its-information-war-taiwan

Drezner, Daniel W., "This Time Is Different: Why U.S. Foreign Policy Will Never Recover," *Foreign Affairs*, Vol. 98, No. 3, May–June 2019, pp. 10–17.

Edel, Charles, and Hal Brands, "The Real Origins of the U.S.-China Cold War," *Foreign Policy*, June 2, 2019. As of March 18, 2020:
https://foreignpolicy.com/2019/06/02/
the-real-origins-of-the-u-s-china-cold-war-big-think-communism/

Edelman, Eric S., *Understanding America's Contested Primacy*, Washington, D.C.: Center for Strategic and Budgetary Assessments, 2010. As of March 18, 2020:
https://csbaonline.org/uploads/documents/
2010.10.21-Understanding-Americas-Contested-Supremacy.pdf

Farrow, Ronan, *War on Peace: The End of Diplomacy and the Decline of American Influence*, New York: W. W. Norton & Company, Inc., 2018.

Finnemore, Martha, *National Interests in International Society*, Ithaca, N.Y.: Cornell University Press, 1996.

Fiske de Gouveia, Philip, with Hester Plumridge, *European Infopolitik: Developing EU Public Diplomacy Strategy*, London: Foreign Policy Centre, 2005.

Flournoy, Michèle, and Shawn Brimley, "The Contested Commons," *Quadrennial Defense Review 2010*, U.S. Department of Defense, 2010. As of March 18, 2020:
http://archive.defense.gov/home/features/2009/0509_qdr/flournoy-article.html

Foundation for the Law of Time, homepage, undated. As of March 18, 2020:
https://lawoftime.org/

Francis, Matthew D. M., "Why the 'Sacred' Is a Better Resource Than 'Religion' for Understanding Terrorism," *Terrorism and Political Violence*, Vol. 28, No. 5, 2016, pp. 912–927. As of March 18, 2020:
https://www.tandfonline.com/doi/full/10.1080/09546553.2014.976625

Freeman, Carla Park, "The Fragile Global Commons in a World in Transition," *SAIS Review of International Affairs*, Vol. 36, No. 1, Winter–Spring 2016, pp. 17–28.

Friedman, Lisa, "E.P.A. to Limit Science Used to Write Public Health Rules," *New York Times*, November 11, 2019. As of March 18, 2020:
https://www.nytimes.com/2019/11/11/climate/epa-science-trump.html

GEF—*See* Global Environment Facility.

Global Commons Alliance, homepage, 2019. As of May 20, 2020:
http://globalcommonsalliance.org/

Global Environment Facility, "Safeguarding the Global Commons," webpage, undated. As of May 20, 2020:
https://www.thegef.org/globalcommons

———, "The Opportunity of the Commons," July 12, 2016. As of March 18, 2020:
http://www.thegef.org/news/opportunity-commons

———, *The Opportunity of the Commons*, June 2017. As of March 18, 2020:
http://www.thegef.org/sites/default/files/publications/
GEF_GlobalCommon%20Quotes%20June2017_r2_0.pdf

———, *Safeguarding the Global Commons: The Seventh Replenishment of the Global Environment Facility*, May 2019a. As of March 18, 2020:
https://www.thegef.org/sites/default/files/publications/GEF_safeguarding_global_commons_May2019_CRA.pdf

———, "Global Commons Alliance to Help Tackle Threats to Health of the Planet," June 6, 2019b. As of March 18, 2020:
http://www.thegef.org/news/global-commons-alliance-help-tackle-threats-health-planet

"Global Thresholds and Allocations Council (GTAC)," webpage, reporting3.org, 2017. As of May 20, 2020:
https://www.r3-0.org/gtac/

Goff, Philip, "Panpsychism Is Crazy, but It's Also Most Probably True," *Aeon*, March 1, 2017. As of March 18, 2020:
https://aeon.co/ideas/panpsychism-is-crazy-but-its-also-most-probably-true

———, "Is the Universe a Conscious Mind?" *Aeon*, February 8, 2018. As of March 18, 2020:
https://aeon.co/essays/cosmopsychism-explains-why-the-universe-is-fine-tuned-for-life

Goldenberg, Alex, and Joel Finkelstein, *Cyber Swarming, Memetic Warfare and Viral Insurgency: How Domestic Militants Organize on Memes to Incite Violent Insurrection and Terror Against Government and Law Enforcement*, Rutgers and the Network Contagion Research Institute, February 2020. As of March 18, 2020:
https://ncri.io/wp-content/uploads/NCRI-White-Paper-Memetic-Warfare.pdf

Goldmann, Kjell, "Realpolitik and Idealpolitik: Interest and Identity in European Foreign Policy," in *European Global Strategy in Theory and Practice: Relevance for the EU*, Stockholm: Swedish Institute of International Affairs, UI Occasional Paper No. 14, February 2013, pp. 6–10. As of March 18, 2020:
https://www.ui.se/globalassets/ui.se-eng/publications/ui-publications/european-global-strategy-in-theory-and-practice-relevance-for-the-eu.pdf

Goodby, James E., "A Global Commons: A Vision Whose Time Has Come," in Sidney D. Drell and George P. Schultz, eds., *Andrei Sakharov: The Conscience of Humanity*, Stanford, Calif.: Hoover Institution Press, 2015a, pp. 131–141.

———, "The Nuclear Dilemma: Constants and Variables in American Strategic Policies," in George P. Shultz and James E. Goodby, eds., *The War That Must Never Be Fought: Dilemmas of Nuclear Deterrence*, Stanford, Calif.: Hoover Institution Press, 2015b, pp. 57–80.

Gorbachev, Mikhail S., *The Coming Century of Peace*, New York: Richardson & Steirman, 1986.

Gouré, Dan, "President-Elect Trump Needs to Prepare to Fight a Massive Information War," *National Interest*, January 6, 2017. As of March 18, 2020:
https://nationalinterest.org/blog/the-buzz/president-elect-trump-needs-prepare-fight-massive-18965

Graham, Thomas, "Let Russia Be Russia: The Case for a More Pragmatic Approach to Moscow," *Foreign Affairs*, Vol. 98, No. 6, November–December 2019, pp. 134–146.

Gray, Colin S., *Hard Power and Soft Power: The Utility of Military Force as an Instrument of Policy in the 21st Century*, Carlisle, Pa.: Strategic Studies Institute, April 2011. As of March 18, 2020:
https://ssi.armywarcollege.edu/hard-power-and-soft-power-the-utility-of-military-force-as-an-instrument-of-policy-in-the-21st-century/

Gregory, E. John, "Control Issues Are Feeding China's 'Discourse Power' Project," *National Interest*, August 15, 2018a. As of March 18, 2020:
https://nationalinterest.org/feature/control-issues-are-feeding-chinas-discourse-power-project-28862

———, "How China Controls Its Citizens," *National Interest*, August 22, 2018b. As of March 18, 2020:
https://nationalinterest.org/feature/how-china-controls-its-citizens-29467

Gross, Terry, "Inside the Russian Disinformation Playbook: Exploit Tension, Sow Chaos," interview with Adam Ellick, *Fresh Air*, NPR, November 15, 2018. As of March 18, 2020:
https://www.npr.org/2018/11/15/668209008/inside-the-russian-disinformation-playbook-exploit-tension-sow-chaos

Grotius, Hugo, *The Freedom of the Seas, or the Right Which Belongs to the Dutch to Take Part in the East Indian Trade*, New York: Oxford University Press, 1916.

Hardin, Garrett, "The Tragedy of the Commons," *Science*, Vol. 162, No. 3859, December 13, 1968, pp. 1243–1248.

Hart, Justin, *Empire of Ideas: The Origins of Public Diplomacy and the Transformation of U.S. Foreign Policy*, New York: Oxford University Press, 2013.

Hartshorn, Max, "Anton Vaino: One Nooscope to Rule Them All," *Mad Scientist Blog*, December 1, 2016. As of March 18, 2020:
http://www.madscientistblog.ca/anton-vaino-one-nooscope-to-rule-them-all/

Herf, Jeffrey, "Russia's Fictional Narratives: A Double-Edged Sword," *American Interest*, December 9, 2019. As of March 18, 2020:
https://www.the-american-interest.com/2019/12/09/
russias-fictional-narratives-a-double-edged-sword/

Heylighen, Francis, and Marta Lenartowicz, "The Global Brain as a Model of the Future Information Society: An Introduction to the Special Issue," *Technological Forecasting and Social Change*, Vol. 114, January 2017, pp. 1–6.

Hochschild, Adam, *King Leopold's Ghost: A Story of Greed, Terror, and Heroism in Colonial Africa*, Boston: Houghton Mifflin, 1998.

———, *Bury the Chains: Prophets and Rebels in the Fight to Free an Empire's Slaves*, Boston: Houghton Mifflin, 2005.

Horgan, John, "Scientific Heretic Rupert Sheldrake on Morphic Fields, Psychic Dogs and Other Mysteries," *Scientific American*, July 14, 2014. As of June 2, 2020:
https://blogs.scientificamerican.com/cross-check/
scientific-heretic-rupert-sheldrake-on-morphic-fields-psychic-dogs-and-other-mysteries/

Hutchens, Michael E., William D. Dries, Jason C. Perdew, Vincent D. Bryant, and Kerry E. Moores, "Joint Concept for Access and Maneuver in the Global Commons: A New Joint Operational Concept," *Joint Force Quarterly*, 1st Quarter, 2017, pp. 134–139.

Hwang, Tim, *Maneuver and Manipulation: On the Military Strategy of Online Information Warfare*, Carlisle Barracks, Pa.: Strategic Studies Institute and U.S. Army War College Press, May 2019. As of March 18, 2020:
https://publications.armywarcollege.edu/pubs/3694.pdf

Ikeshima, Taisaku, "The Notion of Global Commons Under International Law: Recent Uses and Limitations Within a Security and Military Context," *Transcommunication*, Vol. 5-1, Spring 2018, pp. 37–46.

Ishii, Naoko, "We Can Turn the Tragedy of the Global Commons into an Opportunity," *Global Environment Facility* blog, April 4, 2019. As of March 18, 2020:
https://www.thegef.org/blog/we-can-turn-tragedy-global-commons-opportunity

Jasper, Scott, ed., *Securing Freedom in the Global Commons*, Stanford University Press, 2010.

———, *Conflict and Cooperation in the Global Commons: A Comprehensive Approach for International Security*, Washington, D.C.: Georgetown University Press, 2012.

Jasper, Scott, "Maneuver in the Commons," *The Diplomat*, June 10, 2013. As of March 18, 2020:
https://thediplomat.com/2013/06/maneuver-in-the-commons/

Jasper, Scott, and Paul Giarra, "Disruptions in the Commons," in Scott Jasper, ed., *Securing Freedom in the Global Commons*, Stanford University Press, 2010, pp. 1–17.

Jenkins, Tricia, "What Did Russian Trolls Want in 2016? A Closer Look at the Internet Research Agency's Active Measures," *War on the Rocks*, May 22, 2018. As of March 18, 2020:
https://warontherocks.com/2018/05/what-did-russian-trolls-want-during-the-2016-election-a-closer-look-at-the-internet-research-agencys-active-measures/

Kania, Elsa, "The PLA's Latest Strategic Thinking on the Three Warfares," *China Brief*, Vol. 16, No. 13, August 22, 2016.

————, "The Right to Speak: Discourse and Chinese Power," Center for Advanced China Research, November 27, 2018. As of March 18, 2020:
https://www.ccpwatch.org/single-post/2018/11/27/
The-Right-to-Speak-Discourse-and-Chinese-Power

Kavanagh, Jennifer, and Michael D. Rich, *Truth Decay: An Initial Exploration of the Diminishing Role of Facts and Analysis in American Public Life*, Santa Monica, Calif.: RAND Corporation, RR-2314-RC, 2018. As of March 18, 2020:
https://www.rand.org/pubs/research_reports/RR2314.html

Kelly, Kevin, "The Technium and the 7th Kingdom of Life: A Talk with Kevin Kelly," Edge.org, July 18, 2007. As of March 18, 2020:
https://www.edge.org/conversation/kevin_kelly-the-technium-and-the-7th-kingdom-of-life

————, "Two Strands of Connectionsim," *The Technium* blog, January 30, 2009. As of March 18, 2020:
https://kk.org/thetechnium/2009/01/

————, "Protopia," *The Technium* blog, May 19, 2011. As of March 18, 2020:
http://kk.org/thetechnium/protopia/

————, *The Inevitable: Understanding the 12 Technological Forces That Will Shape Our Future*, New York: Penguin Random House LLC, Viking, 2016.

Kissinger, Henry A., *Nuclear Weapons and Foreign Policy*, New York: Routledge, [1957] 2018.

Knopf, Jeffrey W., *Domestic Society and International Cooperation: The Impact of Protest on US Arms Control Policy*, Cambridge University Press, 1998.

Kober, Stanley, "Idealpolitik," *Foreign Policy*, No. 79, Summer 1990, pp. 3–24.

Kofman, Michael, "Raiding and International Brigandry: Russia's Strategy for Great Power Competition," *War on the Rocks*, June 14, 2018. As of March 18, 2020:
https://warontherocks.com/2018/06/
raiding-and-international-brigandry-russias-strategy-for-great-power-competition/

Krastev, Ivan, and Leonard Benardo, "Idealpolitik vs. Realpolitik: The Foreign Policy Debate We Need," *American Interest*, Vol. 15, No. 4, January 8, 2020. As of March 18, 2020:
https://www.the-american-interest.com/2020/01/08/the-foreign-policy-debate-we-need/

Kreisberg, Jennifer Cobb, "A Globe, Clothing Itself with a Brain," *Wired*, June 1, 1995. As of March 18, 2020:
https://www.wired.com/1995/06/teilhard/

Krepinevich, Andrew F., "The Pentagon's Wasting Assets: The Eroding Foundations of American Power," *Foreign Affairs*, Vol. 88, No. 4, July–August 2009, pp. 18–33.

Lakoff, George, *Don't Think of an Elephant! Know Your Values and Frame the Debate*, White River Junction, Vt.: Chelsea Green Publishing, 2014.

Lalwani, Sameer, and Joshua Shifrinson, *Whither Command of the Commons? Choosing Security over Control*, New America Foundation, MIT Political Science Department Research Paper No. 2013-15, April 1, 2013. As of March 18, 2020:
https://ssrn.com/abstract=2256101

LaRouche PAC, homepage, undated. As of March 18, 2020:
https://larouchepac.com/

La Shier, Brian, and James Stanish, "The National Security Impacts of Climate Change," issue brief, *Environmental and Energy Study Institute*, December 20, 2017. As of March 18, 2020:
http://www.eesi.org/papers/view/issue-brief-the-national-security-impacts-of-climate-change

Lazzarato, Maurizio, "The Concepts of Life and the Living in the Societies of Control," in Martin Fuglsang and Bent Meier Sorensen, eds., *Deleuze and the Social*, Edinburgh University Press, 2006.

Lean, Geoffrey, "The Opportunity of the Commons," *Global Environment Facility*, July 12, 2016. As of March 18, 2020:
http://www.thegef.org/news/opportunity-commons

Lepore, Jill, "A New Americanism: Why a Nation Needs a National Story," *Foreign Affairs*, Vol. 98, No. 2, March–April 2019, pp. 10–19. Available at:
https://www.foreignaffairs.com/articles/united-states/2019-02-05/
new-americanism-nationalism-jill-lepore

Le Roy, Edouard, "The Origins of Humanity and the Evolution of Mind," in Paul R. Samson and David Pitt, eds., *The Biosphere and Noosphere Reader: Global Environment, Society and Change*, New York: Routledge, 1999, pp. 60–69.

Levin, Kelly, and Manish Bapna, "Adapting the Commons," *Our Planet*, September 2011, pp. 30–31. As of March 18, 2020:
https://wedocs.unep.org/bitstream/handle/20.500.11822/8041/
-Our%20Planet_%20GLOBAL%20COMMONS%20%20The%20planet%20we%20share-
20111059.pdf

Li Yan, *The Global Commons and the Reconstruction of Sino–U.S. Military Relations*, Asia Paper, Institute for Security and Development Policy, March 2012. As of March 18, 2020:
http://isdp.eu/content/uploads/publications/2012_li-yan_the-global-commons.pdf

Li, Eric X., "The Rise and Fall of Soft Power," *Foreign Policy*, August 20, 2018. As of March 18, 2020:
https://foreignpolicy.com/2018/08/20/the-rise-and-fall-of-soft-power/

Li, Cheng, and Lucy Xu, "Chinese Enthusiasm and American Cynicism over the 'New Type of Great Power Relations,'" Brookings Institution, December 4, 2014. As of March 18, 2020:
https://www.brookings.edu/opinions/
chinese-enthusiasm-and-american-cynicism-over-the-new-type-of-great-power-relations/

Lia, Brynjar, *Architect of Global Jihad: The Life of Al-Qaeda Strategist Abu Mus-ab Al-Suri*, Oxford University Press, 2009.

Libicki, Martin C., "Why Cyber War Will Not and Should Not Have Its Grand Strategist," *Strategic Studies Quarterly*, Vol. 8, No. 1, Spring 2014, pp. 23–39.

Limberg, Peter N., and Conor Barnes, "The Memetic Tribes of Culture War 2.0," *Medium*, September 13, 2018. As of March 18, 2020:
https://medium.com/s/world-wide-wtf/memetic-tribes-and-culture-war-2-0-14705c43f6bb

Lucas, Edward, *Firming Up Democracy's Soft Underbelly: Authoritarian Influence and Media Vulnerability*, Washington, D.C.: National Endowment for Democracy, International Forum for Democratic Studies, February 2020. As of March 18, 2020:
https://www.ned.org/wp-content/uploads/2020/02/
Firming-Up-Democracys-Soft-Underbelly-Authoritarian-Influence-and-Media-Vulnerability-Lucas.
pdf

Lucas, Edward, and Ben Nimmo, *Information Warfare: What Is It and How to Win It?* Washington, D.C.: Center for European Policy Analysis, CEPA InfoWar Paper No. 1, November 2015. As of March 18, 2020:
https://www.stratcomcoe.org/elucas-bnimmo-cepa-infowar-paper-no1-information-warfare-what-it-and-how-win-it

Lynch, Conor, "Did Philosopher Alexander Dugin, aka 'Putin's Brain,' Shape the 2016 Election?" *Salon*, May 5, 2018. As of March 18, 2020:
https://www.salon.com/2018/05/05/
did-philosopher-alexander-dugin-aka-putins-brain-shape-the-2016-election/

Mahan, A. T., "The Influence of Sea Power upon History 1660–1783," in *Roots of Strategy: Book 4*, ed. David Jablonsky, Mechanicsburg, Penn.: Stockpole Books, 1999.

Mardell, Jacob, "The 'Community of Common Destiny' in Xi Jinping's New Era," *The Diplomat*, October 25, 2017. As of March 18, 2020:
https://thediplomat.com/2017/10/the-community-of-common-destiny-in-xi-jinpings-new-era/

Margulis, Lynn, and James E. Lovelock, "Is Mars a Spaceship, Too?" *Natural History*, Vol. 85, June–July 1976, pp. 86–90.

Mattis, Peter, "China's 'Three Warfares' in Perspective," *War on the Rocks*, January 30, 2018. As of March 18, 2020:
https://warontherocks.com/2018/01/chinas-three-warfares-perspective/

Mazarr, Michael J., Ryan Michael Bauer, Abigail Casey, Sarah Anita Heintz, and Luke J. Matthews, *The Emerging Risk of Virtual Societal Warfare: Social Manipulation in a Changing Information Environment*, Santa Monica, Calif.: RAND Corporation, RR-2714-OSD, 2019. As of March 18, 2020:
https://www.rand.org/pubs/research_reports/RR2714.html

Mazarr, Michael J., Abigail Casey, Alyssa Demus, Scott W. Harold, Luke J. Matthews, Nathan Beauchamp-Mustafaga, and James Sladden, *Hostile Social Manipulation: Present Realities and Emerging Trends*, Santa Monica, Calif.: RAND Corporation, RR-2713-OSD, 2019. As of March 18, 2020:
https://www.rand.org/pubs/research_reports/RR2713.html

McCain, Katherine W., "'Nothing as Practical as a Good Theory' Does Lewin's Maxim Still Have Salience in the Applied Social Sciences?" *Proceedings of the Association for Information Science and Technology*, Vol. 52, No. 1, 2015.

McCaul, Michael, "Xi's Long March on American Democracy," *Foreign Policy*, April 5, 2018. As of March 18, 2020:
https://foreignpolicy.com/2018/04/05/xis-long-march-on-american-democracy/

McFaul, Michael, "Russia as It Is: A Grand Strategy for Confronting Putin," *Foreign Affairs*, Vol. 97, No. 4, July–August 2018, pp. 82–91. As of March 18, 2020:
https://www.foreignaffairs.com/articles/russia-fsu/2018-06-14/russia-it

McFate, Montgomery, "The Military Utility of Understanding Adversary Culture," *Joint Force Quarterly*, No. 38, July 2005, pp. 42–48.

McFate, Montgomery, and Steve Fondacaro, "Reflections on the Human Terrain System During the First 4 Years," *PRISM*, Vol. 2, No. 4, September 2011, pp. 63–82.

McLuhan, Marshall, *Understanding Media: The Extensions of Man*, New York: McGraw-Hill, 1964.

McLuhan, Marshall, Quentin Fiore, and Jerome Agel, *The Medium Is the Message: An Inventory of Effects*, New York: Random House, 1967.

Melzer, Scott, *Gun Crusaders: The NRA's Culture War*, New York University Press, 2009.

Meyer, Robinson, "'We Knew They Had Cooked the Books,'" *The Atlantic*, February 12, 2020. As of March 18, 2020:
https://www.theatlantic.com/science/archive/2020/02/
an-inside-account-of-trumps-fuel-economy-debacle/606346/

Michel, Casey, "The Latest Front in Russian Infiltration: America's Right-Wing Homeschooling Movement," *ThinkProgress*, January 17, 2019. As of March 18, 2020:
https://thinkprogress.org/
americas-biggest-right-wing-homeschooling-group-has-been-networking-with-sanctioned-russians-1f2b5b5ad031/

Morgenthau, Hans J., *Politics Among Nations*, New York: Alfred A. Knopf, 1948.

Morris, David, "The Military and the Commons," On the Commons, September 17, 2011. As of March 18, 2020:
http://www.onthecommons.org/military-and-commons#sthash.OaYQtPDE.mAWRlSz9.dpbs

Morson, Gary Saul, "Leninthink," *New Criterion*, Vol. 38, No. 2, October 2019.

Mudde, Cas, *The Far Right Today*, Cambridge: Polity Press, 2019.

Murphy, Tara, "Security Challenges in the 21st Century Global Commons," *Yale Journal of International Affairs*, Vol. 5, No. 2, Spring–Summer 2010, pp. 28–43.

Nakicenovic, Nebojsa, Johan Rockström, Owen Gaffney, and Caroline Zimm, "Global Commons in the Anthropocene: World Development on a Stable and Resilient Planet," Laxenburg, Austria: International Institute for Applied Systems Analysis, WP-16-019, 2016. As of March 18, 2020:
http://pure.iiasa.ac.at/14003/
https://www.iucn.org/sites/dev/files/global_commons_in_the_anthropocene_iiasa_wp-16-019.pdf

National Consortium for the Study of Terrorism and Responses to Terrorism, homepage, University of Maryland, undated. As of June 5, 2020:
https://www.start.umd.edu/

NATO—*See* North Atlantic Treaty Organization.

Navarro, Peter, "The Trump Guide to Diplomacy," *New York Times*, October 15, 2019. As of March 18, 2020:
https://www.nytimes.com/2019/10/15/opinion/trump-universal-postal-union.html

Nelson, Roger D., "The Global Consciousness Project: Is There a Noosphere?" *Golden Thread*, undated. As of March 18, 2020:
http://noosphere.princeton.edu/papers/goldenthread/GTpart4.pdf

Nikonov, Sergey Borisovich, Anna Vitalievna Baichik, Anatoli Stepanovich Puiy, and Nikolai Sergeevich Labush, "Noopolitical Aspect of Information Strategies of States," *International Review of Management and Marketing*, No. 5 (Special Issue), 2015, pp. 121–125.

Nikonov, Sergey Borisovich, Anna Vitalievna Baichik, Rikka Victorovna Zaprudina, Nikolai Sergeevich Labush, and Anna Sergeevna Smolyarova, "Noopolitics and Information Network Systems," *International Review of Management and Marketing*, No. 5 (Special Issue), 2015, pp. 44–48.

Noosphere Forum, Facebook page, last updated June 1, 2020. As of June 2, 2020:
https://www.facebook.com/noosphere/

North Atlantic Treaty Organization Supreme Allied Command Transformation, Alma Mater Studiorum Università de Bologna, and Istituto Affari Internazionali, *Managing Change: NATO's Partnerships and Deterrence in a Globalised World*, June 2011. As of March 18, 2020: http://www.act.nato.int/images/stories/events/2011/managing_change_hr.pdf

Nye, Joseph S., Jr., *Bound to Lead: The Changing Nature of American Power*, New York: Perseus Books, L.L.C., Basic Books, 1990.

———, *Soft Power: The Means to Success in World Politics*, Cambridge, Mass.: PublicAffairs, 2004.

———, "How Sharp Power Threatens Soft Power: The Right and Wrong Ways to Respond to Authoritarian Influence," *Foreign Affairs*, January 24, 2018. As of March 18, 2020: https://www.foreignaffairs.com/articles/china/2018-01-24/ how-sharp-power-threatens-soft-power

———, "No, President Trump: You've Weakened America's Soft Power," *New York Times*, February 25, 2020. As of March 18, 2020: https://www.nytimes.com/2020/02/25/opinion/trump-soft-power.html

Ostrom, Elinor, *Governing the Commons: The Evolution of Institutions for Collective Action*, Cambridge University Press, 1990.

Pace, Scott, "Space Development, Law, and Values," keynote address delivered at the IISL Galloway Space Law Symposium, Washington, D.C., December 13, 2017.

Pakenham, Thomas, *The Scramble for Africa: 1876–1912*, New York: Random House, 1991.

Pamment, James, Howard Nothhaft, Henrik Agardh-Twetman, and Alicia Fjällhed, *Countering Information Influence Activities: The State of the Art*, version 1.4, Karlstad, Sweden: Swedish Civil Contingencies Agency, Lund University, MSB1261, July 2018. As of March 18, 2020: https://www.msb.se/RibData/Filer/pdf/28697.pdf

Pandith, Farah, *How We Win: How Cutting-Edge Entrepreneurs, Political Visionaries, Enlightened Business Leaders, and Social Media Mavens Can Defeat the Extremist Threat*, New York: Custom House, 2019.

Patrikarakos, David, "The Road to (Super)Power," *American Interest*, January 17, 2019. As of March 18, 2020: https://www.the-american-interest.com/2019/01/17/the-road-to-superpower/

Plumer, Brad, and Coral Davenport, "Science Under Attack: How Trump Is Sidelining Researchers and Their Work," *New York Times*, December 28, 2019. As of March 18, 2020: https://www.nytimes.com/2019/12/28/climate/trump-administration-war-on-science.html

Polyakova, Alina, and Daniel Fried, *Democratic Defense Against Disinformation 2.0*, Washington, D.C.: Atlantic Council, Eurasia Center, June 2019. As of March 18, 2020: https://www.atlanticcouncil.org/ wp-content/uploads/2019/06/Democratic_Defense_Against_Disinformation_2.0.pdf

Pomerantsev, Peter, "Inside Putin's Information War," *Politico*, January 4, 2015. As of March 18, 2020: https://www.politico.com/magazine/story/2015/01/putin-russia-tv-113960

———, "To Unreality—and Beyond," *Journal of Design and Science*, No. 6, October 23, 2019.

Pomerantsev, Peter, and Michael Weiss, "The Menace of Unreality: How the Kremlin Weaponizes Information, Culture and Money," *The Interpreter*, November 22, 2014. As of March 18, 2020: https://www.interpretermag.com/the-menace-of-unreality-how-the-kremlin-weaponizes-information-culture-and-money/

Posen, Barry R., "Command of the Commons: The Military Foundation of U.S. Hegemony," *International Security*, Vol. 28, No. 1, Summer 2003, pp. 5–46.

———, "Stability and Change in U.S. Grand Strategy," *Orbis*, Vol. 51, No. 4, Fall 2007, pp. 561–567. As of March 18, 2020:
http://www.comw.org/pda/fulltext/07posen.pdf

Power, Samantha, *The Education of an Idealist: A Memoir*, New York: Dey Street Books, 2019.

Qiao Liang and Wang Xiangsui, *Unrestricted Warfare*, Beijing: PLA Literature and Arts Publishing House, 1999.

Quilligan, James Bernard, "Global Commons Goods: Civil Society as Global Commons Organizations," *Kosmos*, Fall–Winter 2008.

———, "The Commons of Mind, Life and Matter: Toward a Non-Polar Framework for Global Negotiations," *Kosmos*, Spring–Summer 2010.

———, "Why Distinguish Common Goods from Public Goods?" in David Bollier and Silke Helfrich, eds., *The Wealth of the Commons: A World Beyond Market and State*, Amherst, Mass.: Levellers Press, 2012. As of March 18, 2020:
http://wealthofthecommons.org/essay/why-distinguish-common-goods-public-goods

Ramel, Frédéric, *Access to the Global Commons and Grand Strategies: A Shift in Global Interplay*, Paris: Institute for Strategic Research of the Ecole Militaire, Études de l'IRSEM No. 30, October 2014. As of March 18, 2020:
https://www.irsem.fr/data/files/irsem/documents/document/file/1166/Etude_30_En.pdf

Rauch, Jonathan, "The Constitution of Knowledge," *National Affairs*, No. 37, Fall 2018.

Raworth, Kate, "A Safe and Just Space for Humanity: Can We Live Within the Doughnut?" Oxford: Oxfam Discussion Paper, February 2012. As of March 18, 2020:
https://www.oxfam.org/en/research/safe-and-just-space-humanity

———, *Doughnut Economics: Seven Ways to Think Like a 21st-Century Economist*, White River Junction, Vt.: Chelsea Green Publishing, 2017.

Reagan, Ronald, "Address to Members of the British Parliament," Ronald Reagan Presidential Library and Museum, June 8, 1982. As of March 18, 2020:
https://www.reaganlibrary.gov/research/speeches/60882a

Redden, Mark E., and Michael P. Hughes, "Defense Planning Paradigms and the Global Commons," *Joint Force Quarterly*, No. 1, January 2011, pp. 61–66.

Revkin, Andrew C., "Building a 'Knowosphere,' One Cable and Campus at a Time," *Dot Earth* blog, January 4, 2012. As of March 18, 2020:
https://dotearth.blogs.nytimes.com/2012/01/04/welcome-to-the-knowosphere/

Rispoli, Giulia, and Jacques Grinevald, "Vladimir Vernadsky and the Co-Evolution of the Biosphere, the Noosphere, and the Technosphere," *Technosphere Magazine*, June 20, 2018. As of March 18, 2020:
https://technosphere-magazine.hkw.de/p/Vladimir-Vernadsky-and-the-Co-evolution-of-the-Biosphere-the-Noosphere-and-the-Technosphere-nuJGbW9KPxrREPxXxz95hr

Roa, Carlos, "Time to Restore the U.S. Information Agency," *National Interest*, February 20, 2017. As of March 18, 2020:
https://nationalinterest.org/feature/time-restore-the-us-information-agency-19501

Robb, John, *Brave New War: The Next Stage of Terrorism and the End of Globalization*, Hoboken, N.J.: John Wiley and Sons, 2007.

Robinson, Linda, Todd C. Helmus, Raphael S. Cohen, Alireza Nader, Andrew Radin, Madeline Magnuson, and Katya Migacheva, *Modern Political Warfare: Current Practices and Possible Responses*, Santa Monica, Calif.: RAND Corporation, RR-1772-A, 2018. As of March 18, 2020:
https://www.rand.org/pubs/research_reports/RR1772.html

Rockström, Johan, "Common Boundaries," *Our Planet*, September 2011, pp. 20–21. As of March 18, 2020:
https://wedocs.unep.org/bitstream/handle/20.500.11822/8041/
-Our%20Planet_%20GLOBAL%20COMMONS%20%20The%20planet%20we%20share-
20111059.pdf

———, "Planetary Stewardship in the Anthropocene," *Remodelling Global Cooperation: Global Challenges Quarterly Risk Report*, November 2016, pp. 118–123. As of March 18, 2020:
https://globalchallenges.org/wp-content/uploads/Global-Challenges-Quarterly-Risk-Report-
November-2016.pdf

———, "Managing the Global Commons," *Our Planet*, September 2017. As of March 18, 2020:
https://www.thegef.org/publications/
un-environment-our-planet-magazine-september-2017-global-environment-facility-issue

Rockström, Johan, and Kate Raworth, "Planetary Boundaries and Human Prosperity," Inquirer.net, May 1, 2015. As of March 18, 2020:
https://opinion.inquirer.net/84551/planetary-boundaries-and-human-prosperity

Rockström, Johan, Will Steffen, Kevin Noone, Åsa Persson, F. Stuart Chapin, III, Eric F. Lambin, Timothy M. Lenton, Marten Scheffer, Carl Folke, Hans Joachim Schellnhuber, Björn Nykvist, Cynthia A. de Wit, Terry Hughes, Sander van der Leeuw, Henning Rodhe, Sverker Sörlin, Peter K. Snyder, Robert Constanza, Uno Svedin, Malin Falkenmark, Louise Karlberg, Robert W. Corell, Victoria J. Fabry, James Hansen, Brian Walker, Diana Liverman, Katherine Richardson, Paul Crutzen, and Jonathan A. Foley, "A Safe Operating Space for Humanity," *Nature*, Vol. 461, September 24, 2009a, pp. 472–475.

Rockström, Johan, Will Steffen, Kevin Noone, Åsa Persson, F. Stuart III Chapin, Eric Lambin, Timothy M. Lenton, Marten Scheffer, Carl Folke, Hans Joachim Schellnhuber, Björn Nykvist, Cynthia A. de Wit, Terry Hughes, Sander van der Leeuw, Henning Rodhe, Sverker Sörlin, Peter K. Snyder, Robert Constanza, Uno Svedin, Malin Falkenmark, Louise Karlberg, Robert W. Corell, Victoria J. Fabry, James Hansen, Brian Walker, Diana Liverman, Katherine Richardson, Paul Crutzen, and Jonathan Foley, "Planetary Boundaries: Exploring the Safe Operating Space for Humanity," *Ecology and Society*, Vol. 14, No. 2, December 2009b.

Romerstein, Herbert, "Soviet Active Measures and Propaganda: 'New Thinking' and Influence Activities in the Gorbachev Era," in Janos Radvanyi, ed., *Psychological Operations and Political Warfare in Long-Term Strategic Planning*, New York: Praeger Publishers, 1990, pp. 36–68.

Ronfeldt, David, *Tribes, Institutions, Markets, Networks: A Framework About Societal Evolution*, Santa Monica, Calif.: RAND Corporation, P-7967, 1996. As of March 18, 2020:
https://www.rand.org/pubs/papers/P7967.html

———, "Al Qaeda and Its Affiliates: A Global Tribe Waging Segmental Warfare," *First Monday*, March 2005. As of March 18, 2020:
https://firstmonday.org/ojs/index.php/fm/article/view/1214

———, "In Search of How Societies Work: Tribes—The First and Forever Form," Santa Monica, Calif.: RAND Corporation, WR-433-RPC, 2007. As of March 18, 2020:
https://www.rand.org/pubs/working_papers/WR433.html

———, "Explaining Social Evolution: Standard Cause-and-Effect vs. TIMN's System Dynamics," *Materials for Two Theories* blog, September 18, 2009. As of March 18, 2020:
http://twotheories.blogspot.com/2009/09/explaining-social-evolution-standard.html

———, "Speculation: Is There an 'Assurance Commons'? Do Societies Depend on It? Should There Be a U.S. Chamber of Commons?" *Materials for Two Theories* blog, December 3, 2012. As of March 18, 2020:
http://twotheories.blogspot.com/2012/12/speculation-is-there-assurance-commons.html

———, "People's Space-Time-Action Orientations: How Minds Perceive, Cultures Work, and Eras Differ," draft, November 2018. As of March 18, 2020:
https://papers.ssrn.com/sol3/papers.cfm?abstract_id=3283477

Ronfeldt, David, and John Arquilla, "The Promise of Noöpolitik," *First Monday*, August 2007. As of March 18, 2020:
http://firstmonday.org/ojs/index.php/fm/article/view/1971/1846

Rose, Gideon, "The Fourth Founding: The United States and the Liberal Order," *Foreign Affairs*, Vol. 98, No. 1, January–February 2019, pp. 10–21.

Rosenbach, Eric, and Katherine Mansted, *The Geopolitics of Information*, Cambridge, Mass.: Belfer Center for Science and International Affairs, Harvard Kennedy School, May 2019. As of March 18, 2020:
https://www.belfercenter.org/sites/default/files/2019-08/GeopoliticsInformation.pdf

Rothkopf, David J., "Cyberpolitik: The Changing Nature of Power in the Information Age," *Journal of International Affairs*, Vol. 51, No. 2, Spring 1998, pp. 325–359.

Samson, Paul R., and David Pitt, eds., *The Biosphere and Noosphere Reader: Global Environment, Society and Change*, New York: Routledge, 1999.

Scherhorn, Gerhard, "Transforming Global Resources into Commons," in David Bollier and Silke Helfrich, eds., *The Wealth of the Commons: A World Beyond Market and State*, Amherst, Mass.: Levellers Press, 2012.

Schoen, Fletcher, and Christopher J. Lamb, *Deception, Disinformation, and Strategic Communications: How One Interagency Group Made a Major Difference*, Institute for National Strategic Studies Strategic Perspectives, No. 11, Washington, D.C.: National Defense University Press, June 2012. As of March 18, 2020:
https://ndupress.ndu.edu/Portals/68/Documents/stratperspective/inss/Strategic-Perspectives-11.pdf

Shermer, Michael, "When It Comes to AI, Think Protopia, Not Utopia or Dystopia," Edge.org, 2015. As of March 18, 2020:
https://www.edge.org/response-detail/26062

Shiller, Robert J., "Narrative Economics," Cambridge, Mass.: National Bureau of Economic Research, NBER Working Paper No. 23075, January 2017. As of March 18, 2020:
https://ssrn.com/abstract=2903742

———, *Narrative Economics: How Stories Go Viral and Drive Major Economic Events*, Princeton University Press, 2019.

Shirky, Clay, *Here Comes Everybody: The Power of Organizing Without Organizations*, New York: Penguin Press, 2008.

Shultz, George P., "A World Awash in Change," RealClearPolitics, July 12, 2016. As of March 18, 2020:
https://www.realclearpolitics.com/articles/2016/07/12/
a_world_awash_in_change_131161.html

Sideris, Lisa H., "Surviving the Anthropocene Part 2: Of Omega Points and Oil," *Inhabiting the Anthropocene* blog, July 8, 2016. As of March 18, 2020:
https://inhabitingtheanthropocene.com/2016/07/08/
surviving-the-anthropocene-part-2-of-omega-points-and-oil/

Singer, P. W., and Emerson T. Brooking, *LikeWar: The Weaponization of Social Media*, New York: Houghton Mifflin Harcourt Publishing Company, 2018.

Snow, Nancy, and Nicholas J. Cull, eds., *Routledge Handbook of Public Diplomacy*, 2nd ed., New York: Routledge, 2020.

Snow, Nancy, and Philip M. Taylor, eds., *Routledge Handbook of Public Diplomacy*, New York: Routledge, 2009.

Stanley, Patrick, "A Nooscope for World Domination ;-)," *Medium*, August 24, 2016. As of March 18, 2020:
https://medium.com/@PatrickWStanley/
anton-vaino-vayno-vladimir-putins-newly-appointed-chief-of-staff-wrote-a-pretty-far-out-585e90cfaec4

"State Department Officials on Public Diplomacy," video, C-SPAN, September 4, 2019. As of June 2, 2020:
https://www.c-span.org/video/?463937-1/state-department-officials-public-diplomacy

Stavins, Robert N., "The Problem of the Commons: Still Unsettled After 100 Years," *American Economic Review*, Vol. 101, No. 1, February 2011, pp. 81–108.

Steer, Cassandra, "Global Commons, Cosmic Commons: Implications of Military and Security Uses of Outer Space," *Georgetown Journal of International Affairs*, Vol. 18, No. 1, Winter–Spring 2017, pp. 9–16.

Stoessinger, John G., *The Might of Nations: World Politics in Our Time*, New York: Random House, 1965.

Swaine, Michael D., "A Counterproductive Cold War with China: Washington's 'Free and Open Indo-Pacific' Strategy Will Make Asia Less Open and Less Free," *Foreign Affairs*, March 2, 2018. As of March 18, 2020:
https://www.foreignaffairs.com/articles/china/2018-03-02/counterproductive-cold-war-china

———, "A Relationship Under Extreme Duress: U.S.-China Relations at a Crossroads," Carnegie Endowment for International Peace, January 16, 2019. As of March 18, 2020:
https://carnegieendowment.org/2019/01/16/
relationship-under-extreme-duress-u.s.-china-relations-at-crossroads-pub-78159

Teilhard de Chardin, Pierre, *The Future of Man*, trans. Norman Denny, New York: Doubleday, Image Books, [1959] 1964. As of March 18, 2020:
https://ia800703.us.archive.org/1/items/TheFutureOfMan/Future_of_Man.pdf

———, *The Phenomenon of Man*, trans. Bernard Wall, New York: Harper and Row, [1955] 1965.

———, "The Antiquity and World Expansion of Human Culture," as reprinted in Paul R. Samson and David Pitt, eds., *The Biosphere and Noosphere Reader: Global Environment, Society and Change*, Routledge, [1956] 1999, pp. 70–79. As of June 1, 2020:
https://epdf.pub/the-biosphere-and-noosphere-reader-global-environment-society-and-change.html

Terranova, Tiziana, "Futurepublic: On Information Warfare, Bio-Racism and Hegemony as Noopolitics," *Theory, Culture & Society*, Vol. 24, No. 3, May 2007, pp. 125–145.

Thomas, Timothy L., "Russia's Reflexive Control Theory and the Military," *Journal of Slavic Military Studies*, Vol. 17, 2004, pp. 237–256.

Tippett, Krista, "Ursula King, Andrew Revkin, and David Sloan Wilson: Teilhard de Chardin's 'Planetary Mind' and Our Spiritual Evolution," *On Being*, last updated January 23, 2014. As of March 18, 2020:
https://onbeing.org/programs/
ursula-king-andrew-revkin-and-david-sloan-wilson-teilhard-de-chardins-planetary-mind-and-our-spiritual-evolution/

Treverton, Gregory F., Erik Nemeth, and Sinduja Srinivasan, *Threats Without Threateners? Exploring Intersections of Threats to the Global Commons and National Security*, Santa Monica, Calif.: RAND Corporation, OP-360-SGTF, 2012. As of March 18, 2020:
https://www.rand.org/pubs/occasional_papers/OP360.html

Trump, Donald J., *National Security Strategy of the United States of America*, White House, December 2017.

Trump, Donald J., with Tony Schwartz, *The Art of the Deal*, New York: Random House, 1987.

UNEP—*See* United Nations Environment Programme.

United Nations Environment Programme, homepage, undated. As of May 20, 2020:
https://www.unenvironment.org/

United Nations System Task Team on the Post-2015 UN Development Agenda, *Global Governance and Governance of the Global Commons in the Global Partnership for Development Beyond 2015*, January 2013. As of March 18, 2020:
http://www.un.org/en/development/desa/policy/untaskteam_undf/thinkpieces/24_thinkpiece_global_governance.pdf

Uppsala Conflict Data Program, homepage, Uppsala University Department of Peace and Conflict Research, undated. As of June 5, 2020:
https://ucdp.uu.se/

U.S. Army Training and Doctrine Command (TRADOC G-2), *The Operational Environment and the Changing Character of Future Warfare*, July 19, 2017. As of March 18, 2020:
https://community.apan.org/wg/tradoc-g2/mad-scientist/m/
visualizing-multi-domain-battle-2030-2050/200203

U.S. Department of Defense, *National Defense Strategy*, June 2008. As of May 20, 2020:
https://archive.defense.gov/pubs/2008NationalDefenseStrategy.pdf

———, *Quadrennial Defense Review Report*, February 2010. As of May 20, 2020:
https://archive.defense.gov/qdr/QDR%20as%20of%2029JAN10%201600.pdf

———, Joint Chiefs of Staff, *Joint Operating Environment (JOE) 2035: The Joint Force in a Contested and Disordered World*, Washington, D.C., July 14, 2016. As of March 18, 2020:
http://www.jcs.mil/Portals/36/Documents/Doctrine/concepts/
joe_2035_july16.pdf?ver=2017-12-28-162059-917

———, *Summary of the 2018 National Defense Strategy of the United States of America: Sharpening the American Military's Competitive Edge*, Washington, D.C., 2018. As of March 18, 2020:
https://www.defense.gov/Portals/1/Documents/pubs/
2018-National-Defense-Strategy-Summary.pdf

———, *Indo-Pacific Strategy Report: Preparedness, Partnerships, and Promoting a Networked Region*, Washington, D.C., June 1, 2019. As of March 18, 2020:
https://media.defense.gov/2019/Jul/01/2002152311/-1/-1/1/
DEPARTMENT-OF-DEFENSE-INDO-PACIFIC-STRATEGY-REPORT-2019.PDF

U.S. House of Representatives, American Space Commerce Free Enterprise Act, Bill 2809, April 25, 2018. As of March 18, 2020:
https://www.congress.gov/bill/115th-congress/house-bill/2809/text

Vernadsky, Vladimir, "The Biosphere and the Noosphere," *American Scientist*, Vol. 33, No. 1, January 1945, pp. 1–12.

———, *Scientific Thought as a Planetary Phenomenon*, Moscow: V.I. Vernadsky Foundation, 1997.

———, *The Transition from the Biosphere to the Noösphere*, trans. William Jones, in *21st Century*, [1938] Spring–Summer 2012, pp. 16–31. As of March 18, 2020:
https://21sci-tech.com/Articles_2012/Spring-Summer_2012/04_Biospere_Noosphere.pdf

Walker, Christopher, and Jessica Ludwig, "The Meaning of Sharp Power: How Authoritarian States Project Influence," *Foreign Affairs*, November 16, 2017a. As of March 18, 2020:
https://www.foreignaffairs.com/articles/china/2017-11-16/meaning-sharp-power

———, "From 'Soft Power' to 'Sharp Power': Rising Authoritarian Influence in the Democratic World," in *Sharp Power: Rising Authoritarian Influence*, Washington, D.C.: National Endowment for Democracy, International Forum for Democratic Studies, December 2017b. As of March 18, 2020:
https://www.ned.org/sharp-power-rising-authoritarian-influence-forum-report/

Wallensteen, Peter, *Understanding Conflict Resolution: War, Peace and the Global System*, London: Sage Publications, 2002.

———, "Four Models of Major Power Politics: Geopolitik, Realpolitik, Idealpolitik and Kapitalpolitik," in Peter Wallensteen, *Peace Research: Theory and Practice*, New York: Routledge, 2013, pp. 33–45.

Walt, Steven M., "A Manifesto for Restrainers," *Responsible Statecraft*, December 4, 2019. As of March 18, 2020:
https://responsiblestatecraft.org/2019/12/04/a-manifesto-for-restrainers/

———, "Grow Up About Dictators, America!" *Foreign Policy*, March 2, 2020. As of March 18, 2020:
https://foreignpolicy.com/2020/03/02/
2020-campaign-sanders-bloomberg-democrats-dictators/

Waltzman, Rand, *The Weaponization of Information: The Need for Cognitive Security*, Santa Monica, Calif.: RAND Corporation, CT-473, 2017. As of March 18, 2020:
https://www.rand.org/pubs/testimonies/CT473.html

Weinberger, Caspar W., "U.S. Defense Strategy," *Foreign Affairs*, Vol. 64, No. 4, Spring 1986, pp. 675–697.

Weiss, Jessica Chen, "A World Safe for Autocracy? China's Rise and the Future of Global Politics," *Foreign Affairs*, Vol. 98, No. 4, July–August 2019, pp. 92–102.

Wendt, Alexander, *Social Theory of International Politics*, New York: Cambridge University Press, 1999.

Wilson, David Sloan, *The Neighborhood Project: Using Evolution to Improve My City, One Block at a Time*, New York: Little, Brown and Company, 2011.

———, "The Phenomenon of Humanity," Center for Humans and Nature, May 24, 2012. As of March 18, 2020:
https://www.humansandnature.org/to-be-human-david-sloan-wilson

———, "The Tragedy of the Commons: How Elinor Ostrom Solved One of Life's Greatest Dilemmas," *Evonomics*, October 29, 2016. As of March 18, 2020:
https://evonomics.com/tragedy-of-the-commons-elinor-ostrom/

———, *This View of Life: Completing the Darwinian Revolution*, New York: Pantheon Books, 2019.

Wilson, David Sloan, Kurt Johnson, Barbara Marx Hubbard, Richard Clugston, Zachary Stein, David Korten, Mac Legerton, Kevin Brabazon, Doug King, Mike Morrell, and Ken Wilber, "Steering Toward the Omega Point: A Roundtable Discussion of Altruism, Evolution, and Spirituality," Evolution Institute, 2015. As of March 18, 2020:
https://evolution-institute.org/
steering-toward-the-omega-point-a-roundtable-discussion-of-altruism-evolution-and-spirituality/

World Commission on Environment and Development, *Our Common Future*, United Nations, 1987. As of May 20, 2020:
http://www.un-documents.net/wced-ocf.htm

Zakaria, Fareed, "The Self-Destruction of American Power: Washington Squandered the Unipolar Moment," *Foreign Affairs*, Vol. 98, No. 4, July–August 2019, pp. 1–16.

———, "The New China Scare: Why America Shouldn't Panic About Its Latest Challenger," *Foreign Affairs*, Vol. 99, No. 1, January–February 2020, pp. 52–69.

Zalman, Amy, "How Power Really Works in the 21st Century: Beyond Soft, Hard & Smart," *The Globalist*, July 17, 2012. As of March 18, 2020:
https://www.theglobalist.com/how-power-really-works-in-the-21st-century-beyond-soft-hard-smart/

———, "Maximizing the Power of Strategic Foresight," *Joint Force Quarterly,* JFQ 95, October 2019, pp. 14–21. As of March 18, 2020:
https://ndupress.ndu.edu/Portals/68/Documents/jfq/jfq-95/jfq-95_14-21_Zalman.pdf

About the Authors

David Ronfeldt, now retired, worked for more than 35 years at the RAND Corporation as a political scientist. His work resulted in new ideas about information-age modes of conflict (cyberwar, netwar, swarming), future security strategy (guarded openness, noopolitik), and social theory (nascent frameworks for analyzing social evolution and social cognition). He has a Ph.D. in political science.

John Arquilla is distinguished professor of defense analysis at the Naval Postgraduate School. Beyond his work with David Ronfeldt, his books include *The Reagan Imprint* (2006), *Insurgents, Raiders, and Bandits* (2011), and *Why the Axis Lost* (2020). He has a Ph.D. in political science.

While at RAND, Ronfeldt and Arquilla coauthored many reports, including *In Athena's Camp: Preparing for Conflict in the Information Age* (1997), *The Zapatista "Social Netwar" in Mexico* (1998), *The Emergence of Noopolitik: Toward an American Information Strategy* (1999), *Swarming and the Future of Conflict* (2000), and *Networks and Netwars: The Future of Terror, Crime, and Militancy* (2001).